Diamond Safari

Cover design by Todd Haack

Photos of Kaieteur Falls by Robert Haack

Diamond Safari

An autobiography:
Donald Haack
(book 2 of a trilogy)

Some names have been changed to protect the identity of the characters

Library of Congress Control Number: 2007923369

ISBN 13: 978-1-59948-174-6

Produced in the United States of America

Pure Heart Press
Main Street Rag Publishing Company
PO BOX 690100
Charlotte, NC 28227-7001
www.MainStreetRag.com

Acknowledgements

Dr. Robert Fulton for excellent editing and technical advice. My wife, Jan for editing, proofing and keeping event sequences in order. Todd Haack for designing front and back cover and for technical advice.

To the memory of my favorite author, John Steinbeck, for his kind invitation to join him at the Calabash Hotel in Grenada where he planted the seed of these books by convincing Jan and me to write books on the era, area and rough diamond business rather than just memoirs for the family.

ABOUT THE AUTHOR

Donald Haack, a diamond broker has been referred to as the "Indiana Jones of the diamond industry." He has diamond experience as a miner, trader, and broker on the national and international level.

Don was born and raised in Milwaukee and West Bend, Wisconsin. He graduated from the University of Wisconsin with a major in Economics and Finance after which he served in the U.S. Marines at Cherry Point, N.C., his first exposure to North Carolina and Charlotte.

Don spent 20 years abroad as manager and consultant of diamond mining companies in British Guiana and Venezuela, a licensed buyer of rough diamonds, and as a pilot with over 5,000 hours of bush flying. He organized and operated "Guyana Wings," an air charter service to remote diamond mining areas of South America. In the 70's, he designed, built, and operated ocean-going excursion boats in Grenada and St. Martin and maintained international gem connections in Europe where he had one of the biggest sales to a royal family.

He has a rare insight of the inner workings of the DeBeers cartel, and many anecdotes concerning his friend, the late Harry Winston, who donated the Hope diamond to the Smithsonian institute.

After two revolutions in third world countries, Don, his wife, Janet, and four children returned to the U.S. in 1981 where they opened, Donald Haack Diamonds in Charlotte, N.C.

Don served as president of the World Trade Association, past chairman of the Better Business Bureau Board of Directors and Chairman and President of the Foreign Trade Zone. He has served as Chairman of the Rotary Foundation, is on the Charlotte Symphony Board of Directors, and is a member of Metrolina Business Council, Chamber of Commerce, NC Wine and Grape Growers Assoc., and NC Writer's Network. He writes articles, is working on a third book, *Diamonds 'neath My Wings,* lectures in the field of world trade and gems, and is designing a hi-speed ocean-going boat.

To my children, Diana, Thomas, Todd and Julie
for their support and willingness to share in the adventures
and put up with my sometime long absences.

Contents

Donald Haack

PROLOGUE

Our arrival in the States from South America bush country was a major change in life style. The Amerindian saying: *Once you have drunk from the rivers and eaten Laba, Senor, you will return*, was not foremost in our thoughts. We didn't know if we would ever return. It would be hard enough to get acclimated to civilization again.

Bush Pilot in Diamond Country, Book 1, described the mining and trading in diamonds during the first few years in the South American interior. This book, *Diamond Safari,* second of a trilogy on British Guiana (now Guyana) describes the safaris we operated for a couple of years. Book 3, *Diamonds 'Neath My Wings* (a work in progress), is a sequel to this book, describing the trials and tribulations of establishing and running our air charter, *Guyana Wings, Ltd,* servicing Guyana's rugged and isolated diamond country.

Included in this Diamond Safari story are a few of the experiences of our adjustment to civilized life in the USA— mundane chapters compared to our original Guyana adventure. I first decided to leave them out. However, my editor and friends were of the consensus that the U.S. chapters provided an insight on how we coped with the change and the reason we returned to South America...beyond the Indian prophecy about *river water* and *laba*.

While we had not intended to go back to South America, we hadn't completely cut the umbilical cord—the rough-diamond buying trips. The deciding factor to return was provided quite accidentally by our neighbor in Elm Grove, Wisconsin, who was trying to help us out of a dilemma.

Donald Haack

1. COMING HOME

On the Pan Am flight back to New York, I knew there were other good reasons why our family was winging our way back to civilization: the frequency of near disasters and the hazardous life that surrounded us. My thoughts drifted back to our diver, Sankar, who just a few weeks ago narrowly escaped with his life.

It was late afternoon on the Mazaruni River. The dredge spewed out tons of gravel from the bowels of the blackened river with Sankar, the diver, guiding the intake nozzle by feel in the inky water 20 feet below. Only water came forth, which happens as the diver repositions the suction head. But it continued. After five minutes, Adolpho, the diver tending Sankar's life-line and air hose, gave a signal jerk to Sankar, indicating a possible plugged up nozzle. Nothing. A second and third signal, finally the emergency three short pulls. Nothing. Something was wrong. Adolpho pulled on the lifeline to bring Sankar up. Wouldn't budge. Quickly he suited up, jumped in, and disappeared amidst a column of bubbles. We waited. Adolpho surfaced, tore off his mask, shouting "Sankar is pinned below a boulder. Can't pull him out. He's alive!"

Word quickly went out to the other dredges. In the next four hours, three dredges came alongside, offering gasoline, divers, food, and drinks. The dredges tied up and rotated position, ready to replace the five divers below who were arranging rocks in every crevice to keep the boulder from moving and crushing Sankar. The boulder measured more the three yards in diameter. Only Sankar's feet stuck out, which he wiggled in response to being touched and squeezed. That was the only sign that he was still alive. Miraculously he wasn't crushed. One of the divers continued to hold his feet to let him know he wasn't alone.

After two agonizing hours of stabilizing the boulder with small rocks, two divers cranked up their dredges and gently began the process of sucking gravel out from under Sankar. It was hoped that maybe just his upper body was pinned. Two hours of suctioning below him and removing large rocks by hand, only enabled the divers to free his legs up to the hips, but he was still trapped beneath the rock. The strenuous work required that fresh divers rotated every hour. Eventually they were able to free his chest and left arm. He squeezed the hand of his rescuer and indicated where to suction to free his other arm. Guided by his hand signals of yes and no, the rescue divers slowly pulled him back, inch by inch. This was critical because if he were pulled too fast it could rip off his face mask, his only source of air.

While this drama was continuing, gasoline was brought in, standby compressors were hooked up, and underwater lights were used to recheck the placement of rocks below the boulder that pinned Sankar.

Seven-and-a-half hours later, they eased a weak Sankar free. He could barely move on his own. They held his mask to his face and slowly raised him to the surface. When they removed his mask, his lips formed a thank you, thank you, over and over. After consuming quarts of water, he had a cigarette and slept for 24 hours. That was not Sankar's best week. His total share of diamonds: $20. But maybe a fortune next week.

Diamonds were there but the earth was not giving them up easily.

Reluctant as I was to leave our home and business in South America, I had the pending satisfaction that I would finally be in the good graces of Jan's mother and father, who up to that time had not forgiven me for taking their daughter into the primitive jungles and raising their two grandchildren there.

Our arrival at Kennedy International was less than auspicious: a combination of several events that altered the homecoming from what should have been a rousing welcome to one quite different. The bulk of our personal belongings were shipped by boat. The unique and rare collection of Amerindian artifacts brought back by Jan and me from our flights to the remote Wai Wai Indian village was irreplaceable and too much of a risk to be shipped. We carried them on the plane.

They were two seven-foot bows, twelve six-foot brightly feathered arrows, a carved wooden stool, a colorful cassava grader, several small bamboo combs decorated with Macaw feathers, women's beaded aprons, bracelets, and arm bands, a man's monkey-skin purse containing make-up for special occasions, and a one-foot tubular section of painted bamboo with a monkey skin cover that concealed ten hand-carved arrow tips dipped in curare. When they hunted, the Wai Wais inserted the poisonous tips into the long smooth bamboo arrows, which had to be constantly tempered over a flame for true flight. After hitting the prey, the curare tips broke, which saved the arrow, and remained imbedded in the animal, affecting the parasympathetic nervous system— breathing and heartbeat. The animal or bird would die within minutes. The tip could then be retrieved.

We planned on staying in Connecticut with Jan's Uncle Hugh and Aunt Dot in their huge old New Canaan family home until we could regroup and plan our next stateside venture. Bill Mills, Jan's father, was so happy at the prospect of his daughter and grandchildren finally returning to the "civilized" U.S. that he flew to Connecticut to join his brother, Hugh, to meet us at the airport. It was early spring.

Our family left British Guiana on a Pan American Airways flight that overnighted in Antigua. At noon we climbed aboard the DC-4 and left that tropical paradise with its balmy breezes and rustling palm trees lining the airport. Three hours into the flight we were notified that the weather in New York had snow flurries from a cold front moving in. The observant and concerned stewardess noticed that our family was the only one without winter clothes and offered us Pan Am blankets to walk from the plane into the airport.

"Take them and drop them off at the Pan Am counter inside. We're going into a real winter blizzard, and you'll need the blankets."

She was right. We stepped out of the plane and the wind and snow hit us as we walked down the ramp. We pulled the blankets around our heads and only left space to peek out. That was how we entered the terminal building.

Meanwhile, Hugh and Bill were inside checking the disembarking passengers. Hugh saw something that caught his attention and caused an impish little-boy smile that curved his ample mustache. He didn't say a word.

"They're almost all out, and I don't see them, Hugh. Did we miss them or maybe they aren't on that flight?"

"Yeah, I think they're there all right. Watch that group coming up the stairs, now. They're just starting to walk on the ramp facing us." There was a prolonged silence. The group of four, wrapped in blankets and carrying bows and arrows, walked Indian fashion— single file—and approached the upper ramp.

"They sure do have some strange-looking characters coming into this country. Look at those primitive Indians with the bows and arrows; they're straight out of the jungle. Wait 'til they see New York. That'll be a shock." Just about that time, the "Indian" in front pulled the blanket from her head and shook her hair. Snow fell off but there was no mistaking who Bill Mills was looking at—his daughter, Janet, followed by Diana, Tom, and me.

"Oh,my gawd! Oh, my gawd!" Bill exclaimed. Then words failed him, Hugh related to us later.

Hugh had a big belly laugh. He came up hugged us all and relieved us of the blankets, bows, and arrows. He handed the blankets to a surprised attendant who fully expected to see Indians. Jan's face had a question on it.

"Your father is here. He bolted out the door in the opposite direction to wait at the baggage level to compose himself, I guess.

You know how straight-laced he is, Janet. The shock of discovering that the four Indians walking single file, wrapped in blankets, and carrying bows and arrows were his daughter and family…well, it was just too much for his conservative background. I suspect we'll see him in the baggage area when he's a bit less discombobulated. Welcome back to the States," Hugh repeated enthusiastically. "You know, Janet, I was always known in the Mills' clan as the rogue, the non-conformist. That is, until you came along and became the Annie Oakley, jungle bunny, jaguar tamer, and outdid me by light years. I happily concede my title to you. I'm glad you're staying with us, we'll have fun and get you settled in."

Hugh helped guide us through the transition from jungle life to the sophisticated New Canaan, CT society, and introduced us to many of those who commuted to New York City. We found a neat little house in Silvermine, an artist community between Norwalk and New Canaan. Later, Hugh found us an affordable house on Gower Road in New Canaan. He suggested I try my hand at local real estate and introduced me to the Saafs, a family real estate company. It would help keep starvation from our door.

Real estate checks came in but not always in a timely fashion. To augment our income, Jan took a job with Mrs.Bigelow's Constant Comment Tea in Norwalk. She enjoyed Mrs. B., the commercial writing, and the welcome extra dollars.

Amidst all the settling in, we had our third child, Todd, in the Norwalk Hospital. Todd provided a bright spot at a time in our transitional life when we certainly could use it.

Hugh suggested it would be helpful if I joined the local Kiwanis Club. It would give me more contacts in the community. I joined, enjoyed the new relationships, and then was asked to help out on a club project—raising funds to purchase two ambulances for the local hospital. The ambulances were not cheap.

I was assigned a Saturday post with an older man. Traditionally, their annual fund drive began with clowns to attract attention so we put on the uniform of the day, the clown costumes, and stood in front of Gristedes' grocery store, soliciting funds. It was cold but we were dressed warmly. The Gristedes' employees and shoppers were extremely gracious and indulgent, making sure we were warm, had plenty of hot coffee, sandwiches, doughnuts and something of value in our fund-raising pot.

As nice as everyone was, this was not my cup of tea—standing in a clown costume outside a grocery store soliciting money … with another clown, an older sedate gentleman. I wondered how he fit in the picture and if he felt as out-of-place as I did.

In the first part of the morning, while soliciting the shoppers, our conversation was small talk about the people who put dollars in our pot, the local New Canaan news and the need for ambulances. Nothing was said about our individual backgrounds. I wanted to know more about him and by afternoon, we were like old buddies. To keep warm he was stomping his feet and sipping hot coffee more frequently. I had to hand it to him. He was a good thirty years older than me and it had to be a lot harder on him. I was fit and in my twenties. My respect for him went up several notches. I questioned him about his background.

"I work in the computer industry. It's growing, it's fun, and in the next few decades I strongly believe computers will play a major role in the progress of business."

I wanted to know more. Computers were fascinating, and having been out of the country the last few years, I felt out of the technology loop.

"I know your first name, Tom, but I don't know your last. Now that we've frozen here for the day, I feel I should know more about my *compadre in combat.* What company do you work for? Is it here in New Canaan?"

"You're right; we've gone beyond the first name basis. Name's Watson. Company is IBM and it's not too far, over in New York. Well, we not only survived but we collected one helluva pot of dollars in here," he said, changing the subject. "I think you and I have broken the record. Congratulations."

I wasn't listening to his accolades on how we had done. I was still one step behind on the Tom Watson, IBM. I had seen his face before in the Wall Street Journal, *the* Thomas J. Watson, Jr., CEO of IBM, one of the largest of the Fortune 500 companies. The enormity and humor of the situation struck me full force—that I spent the afternoon dressed in a clown suits raising funds for the New Canaan ambulance service with Tom Watson... and now my very humbled feelings. The disdain I felt for this job—compared to one of the most important men in industry who didn't feel it was beneath *his* dignity. Indeed, a humbling experience, and one I would remember the rest of my life.

Selling real estate in New Canaan wasn't exactly my cup of tea, but it did allow me to go to New York, visit some of my college friends, and see what opportunities existed in the big city. I came in contact with a New Jersey business owner who wanted to retire from his caster manufacturing business. No one in his family wanted to take it over. Casters, little rollers below chairs, desks, and industrial benches, did not sound very exciting. The company had a good money-making history and had plans to expand into other fields. The owner, Dan Morris and I spent hours pouring over the company figures, the buildings, and potentials. His bank was willing to extend credit for the transfer of ownership, and there didn't seem to be any obstacles to the take-over.

The family, who for years had not wanted anything to do with the business, suddenly at the last minute had a change of heart and talked Dan out of selling. That was a family affair. I dropped the project.

Meanwhile, back in New Canaan, I sold two residential homes in one month, which brought in some needed income. I didn't enjoy that part of real estate. I perused Saaf's commercial real estate listings and found I had the ability to evaluate solid investments that were overlooked by other brokers.

The first one, a nursing home, was on the books longer than anyone could remember. No one would touch it. I drove out and inspected the thirty-room nursing home, met the owners, staff, and some of the patients. For low-income retirees who needed constant care, it was better than I expected. I went over the books with the owners and was surprised at the impressive cash income it produced. I returned to the office and did a financial *pro forma* on the project to discover it was an excellent tax write-off and returned a high long-term capital gain. After completing the financials, I placed an ad in the Sunday New York Times and received five replies the following Monday. The first prospect bought it two hours after the inspection.

The second property was a bungalow that had been on the market and was, for all purposes, a dead horse as far as the brokers were concerned. No one had showed it in months. The structurally sound house was quaint but run down. Two-foot high grass with other neglected landscaping hinted at abandonment. It needed some tender loving care, which it wasn't receiving.

I checked the town hall records and bingo, as I expected—this little bungalow sat on one side of a double lot, which, on Main Street close to the center of town, should have commanded a premium price.

I called the owner who was disappointed that her house hadn't been shown in the past few months. She was delighted to hear that instead of lowering the price, I suggested we raise it and advertise the second lot or divide it and broker it separately. I asked her permission to have the grass cut and the hedges and low-limbs of the trees trimmed.

The office had a long list of potential customers whom I called. The house and lot were sold the following week—at the higher price with no haggling.

I felt good at my ability to spot the potential of undervalued properties. About that time we had notice of a New Canaan builder declaring bankruptcy. One of his unfinished homes on Woodridge Road, a choice residence, was going on the auction block. I brought in some sub-contractors to price what it would cost to finish the house. Even if their eventual costs doubled their estimates, it would leave a good margin of profit and still be the lowest cost house on the street—always an easy sell.

My neighbor, a school teacher, asked if he could be a partner. I agreed because it made the local financing easier. I won the bid against three other contractors. Two days later we found we couldn't be issued a building permit—the foundation wasn't deep enough for code . . . a slight error of omission. After three sleepless nights anguishing whether to sue to get our money back, I came up with an idea and took it to the building inspectors.

"I've never seen that done before, but it should work. I'll inspect it along the way, and you'll have your building and occupancy permit."

We built a foundation below a foundation, sections at a time. With the deeper and legal foundation, we received our building permit and another good investment.

In spite of the sporadic real estate winners, the payback in real estate was slow and erratic, which didn't make for a happy home life when it came to paying the bills.

At the same time, my brother's visits to New York and New Canaan occurred with unusual regularity. He wanted me to join him in Milwaukee and thought my idea of a design and manufacturing adjunct business to his retail jewelry would be ideal. We would be equal partners. With my rough diamond contacts, we could cut and polish the rough to use in the jewelry. The prospect of a stable income, with the potential to develop our own business, sounded too good to pass. We made the move to Milwaukee. In retrospect, we should have stuck it out New Canaan. We were just getting the hang of commercial real estate.

In Elm Grove, a suburb of Milwaukee, we found a home, fixed it up, and settled in our new business, *Diamond Designs, Inc.*— designing and manufacturing, buying rough diamonds from my sources in South America, and buying and selling unusual large diamonds. That part turned out to be fun...but not necessarily profitable for me.

In Elm Grove my first after-hours project was my attempt at movie making. I called it, *Producing Movies 101*.

I consolidated all my South American 16 mm. films. I bought a book on editing and a two-spool hand editor with all the accompanying paraphernalia. It wasn't long before the shelves in our spare room, now the *library*, were lined with strips of film and a paper describing each scene. Each description was reviewed so that the filmstrips could be listed in logical sequence.

Whenever Jan poked her head in to call me, she took in all the drooping film tapes hung in every conceivable shelf space in the library, shook her head, and rolled her eyes. She didn't say anything but it was all there— *another one of those damnable projects.*

In three months of hard work and many sighs from Jan, I eliminated unacceptable parts and spliced the rest into a twenty-five minute reel of our South American mining stories. I made a copy, which I used for lectures to service groups, museums, private clubs, and anyone who wanted to see and hear about diamond mining in the remote areas of South America. My speaking engagements became a big plus for selling diamonds.

A few months after my first showing, a fellow came into our Majestic Building office to see the Haack brothers. He gave his card to Marcita, our *Gal Friday* who acted as receptionist, hostess, and top salesperson. Jack Douglas of Douglas Productions, Hollywood, California, heard about our film and was interested in placing it on national TV. That got our undivided attention. He said he worked with Animal Kingdom out of Chicago and four companies in Los Angeles who specialized in wildlife and backcountry adventure. We played the film for him as I narrated. It had no soundtrack, but he said that was a plus, not a negative. With my input they could do their own soundtrack with their narrator and proper background music. It sounded good. We made a copy, signed some papers protecting our interest, gave him the copy, and off he went.

Two months later a very excited Jack called to tell us he had lined up a couple of studios. They needed a detailed narrative, some pictures of us, and could we come out immediately? This new venture sounded like fun. We agreed.

Two days later, our taxi dropped us off at 8833 Sunset Boulevard in Hollywood, the office of Jack Douglas Productions, Inc. It was a second story walk-up in a spartan building. The door was open. The interior was slightly more elegant, particularly the receptionist who got up to greet us. She was tall, blond, built like a

Venus with that peach-like complexion that all beautiful California clones have. In short, she was ravishing.

Somehow, I wasn't listening to what she said and neither was Brother Bob. She pointed to the room on the right. "Jack is waiting for you, go right in, and again, welcome to California."

We tried to act normal and stumbled into the room she indicated. Jack greeted us like long-lost friends, talked non-stop, explaining all the projects he was working on and finally came to the point that he made appointments for tomorrow to meet with some producers. They wanted background, a narrative story of the film, pictures they could use for PR, and some photos of us. He scheduled it for eleven the next day.

"No one starts early in this business. Too much partying going on every night. That's where all the business is done," as he gave us a knowing wink and a big smile. I thought it was a bit hammy but this was Hollywood, or at least the lower end of it.

As we were walking out the door and saying our goodbyes, Miss Gorgeous came to the door with us. "Are you doing anything tonight? The grunions are running...and we're all going to the beach for a party. C'mon and join us, what say?"

Her name was Susan. Again paralysis set in while I watched her mouth move. It was sensuous. I was distracted. Bob was worse. Gradually, the thinking process surpassed the emotions.

"Grunions?" What the hell are they? Sounded like something out of an interplanetary invasion. Somehow, I don't think this was the first time this lady produced a zombie-like affect with men. She recognized the symptoms and the quizzical facial expressions.

She answered my unasked question. "Grunions are a small fish about six inches long, something between a sardine and a herring. Once a year at full moon they come by the thousands and wiggle up the beach on each wave, lay their eggs, and then squiggle back down into the water. The point is to catch them as they come up just as they are laying their eggs in the sand. When your bucket's full of grunions, you place them on the barbecue grill with some sauce. In a few minutes, voila, you have a delicacy. Of course they go better with a martini or vodka on ice. You've never done this before? Out here it's a ritual. Wouldn't miss it for anything and *everyone* will be on the beach tonight." She raised her eyebrows waiting for an answer. We dumbly nodded yes. Our ability to talk gradually returned.

"Where do we go? What do we wear? When does it start?"

"Wear *duck*s, shorts, or if you don't have any you can just take off your pants, it's kinda casual. We're going to a friend's house on Malibu Beach. Pick me up here at seven. I have two friends who will come along. Parking is a problem, and it'll save us taking a couple of cars 'cause there'll be lots of people coming. No one misses this event."

After picking up Susan and two girl friends, who were almost as gorgeous, we were directed on a circuitous route that finally wound up in Malibu Beach.

"We're early, so parking is no problem. Hey, Don, I forgot a few things for the party. Want to run down to the store with me and help pick them up?" I was flattered, agreed and off we went. It was a highlight to be shopping with Michelangelo's V*enus.* I had a hard time concentrating on the business of loading up the cart. It was getting full and it occurred to me that this was going to be the *whole* party, not just a few extra forgotten items.

"How are you fixed for money?" she asked as we loaded up a couple more hams. Although the scenery was a bit distracting, I had the foresight (when I first came in and was down one aisle alone) to take out my money clip and remove all but what I thought Bob and I should contribute to this fete. When she asked that anticipated question, I showed her my money clip. She frowned but then quickly took it and said she might have enough to make up the difference. I felt I like the country rube—almost intimidated into financing a party of beach-combers I didn't know. But she gave my arm a squeeze and the thought passed quickly. We arrived back at the beach house where helping hands unloaded the bountiful booty. Graciously, she told her friends that "Don and Bob helped out with the party and be sure to thank them."

The drinking had already taken on a serious tone and guests were in various stages of dress and undress. Missing pants were not an exception. Miss Venus handed me a drink, grabbed my hand, and guided me to the beach. "I'll show you how to go grunion hunting. Take off your shoes and roll up your slacks…or take them off." In the euphoric state I was and holding hands with Miss Sensuous, I thought it would be best to keep my pants on. Who knows how strange I might appear? She was taking charge of me as if I were her property, and I wasn't fighting it. To the contrary—I was in a simple-minded, state of devoted submission. We arrived on the sand. She chug-a-lugged her drink. "Drink up, you'll need two hands." *More than that if one of them is holding yours*, I thought. Thinking clearly was not one of my abilities at that time. We picked up two small buckets and ran down to the water's edge, hand-in-hand. This was exciting.

"Okay, here they are." We followed up the wave as these squiggly little things were going up with us. She dropped my hand and grabbed fish, two, three, four at a time, placed them in the bucket and grabbed more.

"C'mon, you'll never fill your bucket that way. Grab in front of them and they'll go right into your hand." She was right and I quickly filled my bucket. I was so caught up in the hunt that I hadn't noticed that Miss Venus had disappeared. I went to the barbecue where another girl helped me spread the grunions on the grill, as she brushed them with a sauce.

"You did a good job," this gorgeous clone confirmed as she brushed my cheek with a kiss. "You need a drink to help wash these down. She returned with a vodka and ice, a paper plate with a couple of buns, and placed four of the grunions on the buns. She took a bite and a sip of the drink.

"Salud. Now it's your turn, and she held the sandwich for me to bite. I did. Then the glass for me to drink. I did. *Handy and accommodating these California clones.* In a minute she, too, was gone. I wondered if they were real or figments of my imagination. I glanced around. Brother Bob was in full swing of the routine. He had one hand around his Venus clone, his face in a three-drink beatific smile. These parties kind of get to you.

Donald Haack

I wandered back into the house to see what was going on there. At the far side of the room there was a gathering of people, and I could see the center of attraction, *my Venus,* Susan. I came up to the circle. She had stopped talking and another blond beauty related a story that had everyone laughing. Susan held an ice cream cone and was licking it as she rotated the cone. I wasn't the only one watching her. It was a ritual, an obscenely salacious, sensual one. In a lifetime, I had seen many people lick ice cream cones, but this was something special. Maybe it was because she was special. As I was transfixed in my stare, a piece of the ice cream slid off the cone and dripped down her long tanned leg. It just missed her short shorts. As it melted, it trailed down to her knee.

It was like watching an "X" rated movie, only it was real and getting better, or worse, depending on who were the involved characters. My first reaction was to find a napkin and hand it to my fair lady. Before I could react, a scruffy character in torn shorts and a stained tank top, came up to her, dropped to his knees, sliding into a frontal attack position. The "X" movie progressed. One hand was behind her knee, the other on her thigh at the shorts level. He bent over as if to kiss her thigh but instead started licking off the ice cream, starting from the knee and slowly, deliberately going up the thigh. He had a wicked smile, half-closed eyes, and made groans that were a good imitation of an orgasm, which he may well have been having. All this taking place with *my* Venus.

If you've ever run out of a sauna in the winter and jumped and rolled into a snow bank and then back into the hot pool, you'll understand the feeling I was experiencing. It took a few moments for me to get my equilibrium and understand that this particular goddess was not *mine.* That was a temporary illusion or delusion, a pretty heady one at that. But here in Hollywood I figured this kind of thing goes on all the time. I chuckled at my blind-sided, Midwestern, naïve attitude.

I was still holding my drink as I sidled off to a less crowded part of the beach house. There was a lone guy sitting on the floor leaning against the wall, taking in the whole scene with a half-cynical expression displayed on his bearded face. He might be interesting to talk to. I walked over and asked if he minded if I joined him.

"Please," as he extended his hand for me to join him and hitched over slightly. He didn't need to. We had the whole wall to ourselves. Leaning with my back on the wall, I slid down without spilling my drink, which I thought was pretty good for this time of the fete.

"Don," I said. "Joel," he answered.

I wasn't sure if I was imposing on this man's quiet reverie and hesitated to start a conversation. He initiated a camaraderie by clinking my glass in a toast.

"Here's to parties, may they go on forever, or at least seem like they do. I wonder if this crowd knows how to do anything else with such enthusiasm." He placed himself outside of this group and assumed I was also. Perhaps, I was thinking, Midwestern conservatism stands out like a sore thumb.

You're not part of this feting crowd then?" I asked.

"Yes and no. I'm not part of this crowd, but they come to my house, so in a way I am a regular, if not a participant."

"This is your house?"

"My wife's and mine. She first wanted it for a get-away, but it turned out just the opposite. There are parties all the time but she's seldom here. She invites friends who have friends... and they all come. How did you happen to be here? You don't seem to be a regular in this party crowd."

I explained my situation and pointed out my brother who blended in with the crowd better than I did. "Is your wife here now?" I asked.

"No. My wife is Claudette Colbert, and she likes her privacy too much to come here when the partying is going on. I'm here to keep an eye on things and to make sure the clean-up people do their job in the morning.

Claudette Colbert, I mused. One of the highest-paid female actresses in the movie business... and she can't enjoy the privacy of her own beach house.

Donald Haack

The Midwestern contingent pooped out early. We gave our thanks to faces that couldn't understand our leaving so early— it wasn't even light yet.

The next morning we finished all our business with the producer and editors and were assured they could place our film on several programs. Susan, whom I'm sure never slept, looked as chipper and bright as ever. She gave us a goodbye hug. We left that afternoon.

There wasn't much communication from the California office. We heard from friends in Chicago that *Animal Kingdom* ran our film. Our cousins in San Diego called to say they saw our film on an LA show. Our calls to Jack went unanswered. A girl, not Susan, assured us we would hear from him shortly. He was busy traveling and booking shows. Our frequent calls to his office produced no results. Six months passed. His wife called to tell us Jack died. We extended our sympathies. She had no idea where our film was. Two years later our film arrived at my brother's office. No note, no explanation, just the film. End of story. End of film-making 101.

2. BUYING TRIP

The contacts we made from the years we lived in British Guiana were invaluable, so I didn't want to lose them from disuse. A problem I encountered on my last trip by commercial airlines was the inability to find a small charter aircraft that would fly into my old Guiana bush strips.

The alternative was to fly down in a small plane that I could use in the interior. I thought about it and unexpectedly had the answer the following day— a phone call from a Mrs.Thoger Jungerson. A more interesting couple than the Jungersons would be hard to find. They were naturalized Americans originally from Denmark. I'd met them through their son, Ted Thoger, Jr., who toward the end of our stay in B.G., flew his own plane to the Monkey Mountain airstrip. He was in the process of building a house and airstrip ten miles east along the Echilebar River.

Thoger, Sr. was a brilliant designer. He had the patent for the quick-release seat belt used in airplanes and a patent for welding blades onto jet aircraft that reduced labor time and costs and increased the quality and efficiency of the engine. His latest *toy* was a lightweight aluminum diamond-separating jig that he wanted to test before selling it to DeBeers or other diamond mining companies. Testing the jig is what brought him to the diamond-bearing Echilebar area.

"Don, this is Helga Jungerson, Thoger's wife. I have a favor to ask. Are you planning on a trip to B.G. any time soon?"

I started to tell her what I was planning but she interrupted.

"The reason I ask, Thoger is in the hospital where he is undergoing an appendectomy. He planned to fly down to meet Ted when he had this attack. The doctor doesn't think there will be any complications, but Thoger "the invincible" will be leaving the hospital in two days and then has plans to fly our Bonanza to South America by himself. You've heard about his flying episodes and understand why I don't want him flying by himself anymore, particularly not right after a major operation. I can't go with him. I have a full schedule of meetings I'm not able to cancel." There was a pause waiting for my answer.

"I was planning a buying trip in the next couple of weeks but hadn't firmed up the date, yet."

"I can get Thoger to postpone a couple of days. He would enjoy your company. Could you move your date earlier? I would really appreciate anything you could do. You know all expenses would be paid."

And so the plans were made. I arranged the necessary financing the next day for the diamond buying, which delighted my brother, Bob, who could then buy diamonds right from the mine at a price he couldn't get from regular suppliers. Four days later I arrived at Jungerson's home in Summit, NJ. No time was wasted. I arrived by taxi and Thoger, who was at the door waving, told me to hold the taxi.

"We'll use him to take us to the airport. Helga is out on an errand, and I'll leave her a note."

I chuckled at Thoger's single-minded decisions, most of which never included details. He was leaving for a three-week trip, and was leaving his wife a *note*. He scribbled it, left it on the kitchen table, picked up his hat and bag, and walked out the door. I closed it.

"Don't you want to lock it?" I asked.

"I don't have a key. It'll be all right. Let's go, the cabbie is waiting." And so we left.

Fortunately, there were excellent maintenance facilities at the airport. Thoger didn't skimp when it came to keeping the plane in good shape, a fact reinforced by my very thorough preflight inspection. We packed and took off. Four hours later we approached Fort Lauderdale International Airport and Jungerson picked up the microphone.

"Hello, Ft. Lauderdale. This is Mr. Jungerson, Bonanza 51 Delta coming up on your airport. Do you hear me?"

I flinched. This was not radio protocol, and I expected a reprimand from the tower. I was taken aback by the response.

"Hello, Mr Jungerson. Nice to have you back. Do you have another pilot with you today?"

"Oh, yes. Mr. Haack is a pilot with me in the front seat. We're going to British Guiana and need a fuel stop."

"If you're flying the plane, why don't you let Mr. Haack use the radio. Capt. Haack, are you listening? Can you take over?"

"Yes, I'll be glad to. We're at 8,000 ft., twenty miles out on a 185 heading. ETA Ft. Lauderdale in ten minutes, 1350 local time. Transponder transmitting on 7650. Over." There was an audible sigh of relief from the communication operator when he heard my voice.

"Bonanza 51 Delta, have you on radar and transponder. Proceed and switch to tower frequency ten miles out. Light traffic to your west. None in your area."

I clicked the transmitter twice to acknowledge the message. Mr. Jungerson was well known at this airport. First, because he passed through several times in the past few months, but more so because Jungerson flew from B.G. with his son, daughter-in-law, and four-year-old grandson only two months ago. They were twenty miles southeast of Miami when they ran out of gas and ditched in the ocean. Fortunately, they had just cleared the ATIZ, Air Traffic Identification Zone, and Miami Communication had their exact location on radar. When Jungerson advised his engine had quit, the Coast Guard was on their way and arrived only a few minutes after the aircraft hit the water. Luckily, it was a calm day, the ocean relatively smooth, and the aircraft did not flip over. All four had their life jackets on, and they were settling in the life raft when the plane sank and the Coast Guard arrived. The people in Florida knew Mr. Jungerson intimately.

He was a most remarkable man. A year ago on an invitation from his son, Ted, Jan and I drove down from New Canaan and visited the family, where we heard many of the absent-minded professor tales. Fortunately, his good sense of humor and ability to laugh at himself fueled Mrs. Jungerson into relating some of the more bazaar stories. One he told on himself.

"I had these terrible red blotches on my face. I looked like a freak. In public, people stared at me. The doctors told me I would just have to live with it, but I couldn't. I would become a recluse. So I took some sandpaper, drank part of a bottle of vodka, and then sandpapered my face. With all the blood, it was hard to tell if I was getting the red out or if it was all blood. When I finished, I kept my face moist and sterile with wet cloths boiled in Epsom salt. Three times a day I took mega doses of vitamin C and in three weeks my face cleared up. You can only see a little bit where I must have missed. See? If you look closely you can see my handiwork. When I showed it to the doctors, they couldn't believe it. But it worked. I don't look like a freak anymore."

Mrs. Jungerson explained he really is the *absent-minded-professor* on a regular basis. "Many nights when he comes home, he will sit down, eat dinner, then excuse himself to the den where he studies or is working on another patent. After an hour or so he will come out and ask what we are having for dinner and when will it be? I tell him we had a meal of pork chops, mashed potatoes, and green beans. He isn't surprised and his usual comment is, 'Okay, then I will go back to work.' And he returns to his den."

His son, Ted told me about another flying trip his father made alone. He was to land and overnight in Trinidad before flying the last leg to British Guiana. It was late in the afternoon, the sun was low in the horizon, producing extravagant color changes in the sky, ocean, and the palm tree island. It was too beautiful and peaceful to stop so he decided to continue to B. G. By the time he crossed the Orinoco delta, the sun set and darkness settled in within minutes. Using the reflection of the water, he flew along the coast, reaching Georgetown about eight o'clock where he followed the reflected light of the Demerara River west to Atkinson field.

Donald Haack

With very few flights into Guiana, the airport closed down when there weren't any international flights coming in. Consequently, no lights. When Thoger calculated he should be in the Atkinson area, he circled for a few minutes hoping someone would hear the plane and turn on the runway lights, which is exactly what happened. The Shell Oil people were working late, and when they heard an airplane flying around they understood it was an emergency and called the tower personnel. Everyone raced to get the lights on the field, and within minutes Jungerson's plane touched down.

They cleared him to the main hangar to await recently notified customs and immigration. It would take some time before they would arrive and would he please just park in front of the terminal and await them? When his plane remained on the main runway, they were confused. After several exchanges on the tower frequency, it became evident Jungerson was completely out of gas, not enough to taxi to the hangar. When I heard the story I couldn't help but think God must look after some people who just need looking after. But He must have been a bit put out when it happened a second time when Jungerson didn't quite make Ft Lauderdale and had to ditch in the ocean. God, after all, must have limited patience.

Fortunately the remainder of our trip was uneventful, and Mr. Jungerson and I took turns flying. The time went by quickly while we exchanged stories of our B.G. experiences.

The buying trip was successful. Not too unexpectedly, my brother, who doesn't like to be left out of any adventure, decided at the last minute to fly down commercially and join us on the return trip. He spent three days renewing his acquaintance with Dennis Khouri, the ultimate host and son of the man who started the Khouri Department store, one of the bigger establishments in Georgetown. Our biggest problem was convincing Bob we were actually leaving, and he had to be ready if he wanted to fly back with us. Breaking him away from Dennis Khouri's never-ending hospitality became our biggest obstacle to overcome on the trip. We finally got him on the plane with us and off we went.

The trip was uneventful until San Juan. The weather was scuddy, and we had to make some deviations around a few thundercloud clusters, delaying our arrival for our fuel stop in Puerto Rico. No matter how hard we tried to split the chores of refueling and going through the red-tape procedures of entering U.S. territory, we couldn't complete the process in less than one to two hours. Even with the delay, we had ample daylight flight time to reach the Bahamian island of Great Exuma and home of Morton Salt, for a fuel stop and overnight. The weather report indicated partial overcast but no storms in the area. At 10,000 feet, we would be in the clear and have intermittent holes to see through to descend visually when we arrived. Sounded like an easy flight. Planning and reality are not necessarily the same.

We were 90 minutes out of San Juan. I was flying, Jungerson co-piloting, and Bob in the back seat. I was navigating by dead reckoning: time, distance, and direction. Great Exuma had no beacon or radio station to fix on with my ADF, All Directional Finder. All calculations were double-checked by the signals received from two islands: Grand Turk to the north, the easternmost Bahamian island, and the ADF station on the northeastern tip of Dominican Republic. By drawing a line on the map from each station, the intersection would approximate our position, which at this time indicated we were twenty miles northeast of Dominican Republic. I did the navigation fix and almost immediately we experienced strange happenings.

Jungerson, sitting in the right seat, nudged me and poked his finger towards his side window. Eyebrows raised, face contorted into a big question mark and his head tilted, he indicated I should look out the window. Less than a hundred feet away was a huge, long slender totally black aircraft without markings. We could easily observe the pilot in the cockpit. He was facing us, or more accurately, inspecting us. It was more than a strange feeling—no indication we should pick up the mike or to take any action; he was uncomfortably close and staring at us. The lack of any identifying markings on the plane caused the hair to stand up on the back of my neck. . . unnerving to say the least.

My brother was the first to speak. "Geez, he's close. Who or what do you think it is?" he asked no one in particular.

I answered, "I've seen a lot of planes but nothing like this. Check the wingspan—enormous. Plane's as big as anything I've ever seen, but the body is needle thin…and there isn't a mark or identification on it. What the hell do you think is going on? He's inspecting us, whoever *he* is but not making any signals to us. Kinda spooky." I raised my hand in a half wave to see the reaction I would get. The pilot turned away from his direct gaze on us, and a fraction later the plane was gone. It went up out of my vision at rocket speed. Jungerson moved his head close to the glass to see where the mystery plane went.

"It went straight up, and it's out of sight already. What kind of airplane has that kind of performance?" Jungerson was as confused as we were.

I pulled our plane up at an angle for a better view out the top of the windscreen. He was right—nothing in sight. It was gone in a couple of seconds. Unnerving as it was, I still had to fly our plane, and my attention went back to the instruments and then back to scanning the horizon. I didn't know what I expected to see. San Juan radio hadn't reported any traffic in the area at 10,000 feet, which meant they didn't have an *out-of-space* weird aircraft on their radar either. A coincidence? Or what really was going on?

My reverie of circling thoughts with no logical answers was interrupted by what I saw directly ahead of us. Nothing was making sense and this new apparition added to the mystery—a mountain top straight ahead! I checked my elevation: 10,000 feet.

I asked Jungerson. "Am I seeing what I think I'm seeing out there? A mountain top or am I hallucinating?"

"It's a mountain, and you're not hallucinating. All three of us saw the plane and I can see the mountain." Bob had already confirmed seeing it.

"Yeah, fine, only we're supposed to be 10,000 feet over open ocean at twenty miles northeast of Dominican Republic. So what's with the mountain poking through the clouds? The map shows the closest mountain twenty-five miles *south* of us on Dominican Republic. Either our instruments have suddenly gone haywire or they moved the island."

Donald Haack

The minute I said it, a bell went off. "Or, more logically, Dominican Republic moved their beacon from the northeastern coast to the south side of the island." It's the only answer how we could be twenty- five miles off course. The Republic wouldn't have had any compunction doing this without notifying other air authorities. Grand Turk is too small an island to make any significant difference if they moved their beacon. "Problem is, we're off course and not knowing exactly where the Dominican beacon is, I can't get an accurate position fix. With limited daylight we'd be guessing where Great Exuma is. They don't have a beacon or lights."

After a quick check on our position and a measurement to Great Exuma and Grand Turk, I made a decision. "Gentlemen, we don't have much choice. We have one reliable beacon, Grand Turk. It's the same distance as Great Exuma. But we know exactly where Turks is, and we don't have any way to accurately navigate to Great Exuma. I'll call San Juan and change our flight plan. I have to anyway, because I noticed on the big San Juan map Turk's island is restricted for whatever reason."

"San Juan radio, Bonanza 51 Delta. Do you read me?" We were on the marginal limits of VHF communication with San Juan. They answered but were breaking up badly. I repeated the message and change of plans several times before I received back part of a message— the island is in a restricted zone out of bounds for private aircraft. I transmitted back that this was not a choice, it was an emergency, and we were going in with no other options.

"Please advise them if they can be reached." The rest of the communication was unintelligible, and all I could do was hope most or some part of my transmission got through. I turned the Bonanza onto the new heading, straight to the Turks Island ADF marker. I pulled back the pitch on the prop, set the rate of descent at 300 feet per minute, which at our ETA in thirty-five minutes, should put us at near sea level at the time we sight Turks. With the engine running at full cruise RPM and descending 300 feet-per-minute, we were cruising at just below the red line, the best speed we could hope to squeeze out of this Bonanza. We were racing against darkness. I didn't want to attempt a night landing, and I wasn't sure they would receive our message from San Juan and turn on landing lights for us—if they had any. It was a silent flight for the three of us as we watched the minutes and miles tick by.

In thirty-two minutes it was getting dark and hazy, and I saw the first whitecaps of the water a couple of hundred feet below. Then, the outline of Turks Island. We not only homed into the island with no deviation in direction, but I could make out the white coral strip separated from the darker vegetation around. It was dead ahead, a good feeling. I could land whether airport lights came on or not. As it turned out, they didn't. I used our plane's landing lights to determine height for landing. The white strip showed up clearly in the half dark. I gave thanks for small favors.

The landing was uneventful and smooth. When I turned and taxied back to park near the road we passed over, it was almost dark. As I came closer, I saw the outline of a larger plane parked to the left. Since no one came out to guide us, I parked alongside and shut down the engine. There was an audible sigh of relief from all three of us—safe on terra firma, a good feeling.

When we were landing I noticed some lights to our right, about a mile down the road. Since no car came out to transport us, it might be a good mile walk. I was hoping we could find a place to sleep, eat, and have a much-needed beer.

"Why don't we just take our shaving kits, a shirt, and a change of shorts? We may have a long walk, and I wouldn't want to lug a suitcase and work up more of a sweat." Both Bob and Mr. Jungerson readily agreed. Jungerson was rummaging through his clothes from his case on the ground, Bob was pulling out his suitcase. I was waiting patiently.

"Hey guys, lookee what we have here," and I aimed my flashlight on the plane parked alongside. In the half-light we hadn't taken notice but now standing below and shining a light on it, the plane was enormous. "It's our mystery craft that gave us the once-over at 10,000 feet. Not a single identifying mark on it." I moved the light over the tail, fifteen feet above us.

Donald Haack

Our comments about this strange apparition were interrupted by the sound of a motor and the bouncing lights of a vehicle driving down the airstrip towards us. It approached and stopped. Our flashlights outlined a jeep and three men in Marine fatigues.

"Hey, Bob, they're fellow Marines." I called out *Semper Fi* and waved as they climbed out of the jeep.

"You guys Gyreens?" one of them asked.

"Two of us are—were. . . sergeants. I'm Don; did the Parris Island boot camp. Brother Bob here is a San Diego Marine, but don't hold it against him. What are you guys doing pulling this kind of duty on a paradise island? Oh, excuse me, meet Thoger Jungerson, it's his plane." We all shook hands and heard their first names.

I explained the navigation problem and the apparent screw-up of Dominican Republic moving their station and why we had to come here, even though it was listed as restricted.

"We didn't have much choice with time running out. Is there any place we can get food, beer, and a place to bunk tonight?"

The Marines weren't wearing any rank, but I guess at this time of night they weren't expecting any guests. The tallest one, in charge, did most of the talking.

"C'mon, we'll take you over to our barracks; sign you in because this is a restricted base." He didn't elaborate. "Then we'll find a place in the local settlement where you can get some R & R. Shouldn't take long." We hopped in the jeep and took off in a cloud of reddish coral dust illuminated by the red tail lights.

The barracks were new and temporary but adequate. We had to make out a report on why we landed, how long we were staying, our identification passports, where we lived, purpose of trip, and what color underwear we had, or so it seemed. As we finished the last question, Jim, the tall guy said quietly, "Oh shit, here they come. I thought we might avoid those assholes." I glanced to the door where his attention was focused and saw two men in suits stride in as if they owned the place. And then, as they say, the proverbial shit hit the fan.

Donald Haack

Officious and obnoxious to the extreme, they shouted the same questions we just answered. They threatened us with jail, impounding our aircraft, having our licenses revoked so we would never fly again, and there was probably more I chose to forget. I had met guys like this before. And they were certifiable *assholes,* as the Marine said. I was glad to see he stood alongside me. Every time Mr. Obnoxious closed in, my tall Marine edged in closer, effectively keeping him at more than a smelling distance.

Knowing the training we all went through, this Marine was making sure I didn't lose patience with this idiot and shove his arm down his throat. It was a good feeling to have his kind of backup. I let our newcomer rave until he couldn't think of anything else to accuse and threaten us with. I was about to speak up when I was up-staged by Thoger, who was totally quiet and inconspicuous during this confrontation but now rushed forward in an end run around my Marine protector and right up to Mr. Obnoxious. My Marine friend was about to step in between them when I restrained him. I shook my head and held up my pointed finger in a *wait-and-see* sign.

Jungerson stuck his finger within an inch of the guy's nose. "Are you American or what?" he demanded. I had known Jungerson for a couple of years but had never seen him in such an aggressive and authoritative attitude. He could have been taken for an ambassador or high-ranking politician in the role he had taken on. Jungerson's authoritative manner caused our civilian jerk to step back and re-evaluate the situation. Everyone has to account to someone, and this guy would not like to screw up and be taken to task for taking on the wrong high-ranking individual.

Thoger didn't let up. "You're in civvies and have a military escort. So you're CIA or some other snoop organization. You've threatened us enough. Let me tell you what I intend to do. I'm going to inform my congressman from New Jersey and the Secretary of State, whom I worked with last month, and ask him just what kind of Keystone Cops we have over here. This is supposed to be security? We notified San Juan two hours ago we were coming in on an emergency flight plan. You didn't hear our aircraft approach?

Donald Haack

You took over an hour to finally find us here? Your bodyguards, these Marines here, were on the spot before we even were out of our plane. Makes you look pretty bad, I'd say. You better come up with some good answers, particularly if you're trying to hide the funny plane we're parked alongside."

Me? I wouldn't have said a peep about the *funny* plane. But Jungerson was on a roll…and it worked. Our snoop backed off a bit but had to save face. Jungerson knew enough to let him.

"One. I want you out of this area tonight."

"Two. You're going to be fueled tonight, and you will be off this island before daybreak, 0610 hours.

"Three. I want your word that nothing, and I mean nothing you have seen and heard tonight will be discussed outside with anyone. If you agree, I will not press any charges, and this will not be reported to the FAA." He faced each one of us waiting for our personal verbal agreement. Obviously we agreed. He turned back to Jim, our Marine.

"I can count on you to have them taken off base with a couple of guards to see they do not wander off anywhere else? We can have their plane fueled. Can you bring them back so they can leave before daylight?"

Jim assured him it would be no problem. They turned and left the room.

Jim went up to Jungerson. "Watching the dressing-down you gave him went right to the feel-good spot in my heart. Good for you. And you hit it on the head— they're CIA and the worst assholes in the world." In reference to my comment on the great duty on this paradise island, "This would have been the greatest tour of duty anyone could ask for if it wasn't for the pompous play-acting CIA assholes"— the fourth time he used the endearing term about them.

"You guys must be starving. Bill and Chuck, pack an overnite and take the three of them to the little village north of here. It's the only bar open, and they can get some food and a beer and someone should be able to put them up for the night."

The bar was more like a store. They made some sandwiches with sardines from a tin and a couple slices of onion. We washed it down with cold beer, several of them.

Bill, who had been talking to the bar keep, explained to us there was a shack five minutes walking from here with cots in it, or the barkeep could sling hammocks on the front porch right here. "Did you ever sleep in hammocks?" We assured him we did, and the three of us opted for the hammocks.

It was a short night. We slept fitfully and were awakened by a flashlight shining in our faces. Chuck shook the hammock.

"What time is it?" I asked. It was too dark to see my watch. "Is it morning already?"

"Zero-five-thirty. It'll take us ten minutes to get you back to the airport and I'm sure you want to check your plane out thoroughly before you take off in the dark." He was right. When we arrived at the airstrip, the big plane was gone. I wondered how they could have taken off without us hearing it. Jim was there with his jeep.

"After you left I went back to the plane and stood around while they refueled you. I don't trust these bastards and wanted to make sure they didn't screw up something on your plane. Wouldn't put it past them. After they topped up your tanks and left, I put a guard standing by all night just in case they came back with some strange plans for you. We Marines have to stick together."

While I finished a thorough pre-flight inspection, Jim walked around with me. "They gave instructions for you to be off the ground before day break and out of the area immediately. But just as a precaution, get up to 3,000 feet and circle…in case anything goes wrong you can still ditch back here. Just a precaution, mind you, but I wouldn't put it past them to assure you don't make the next island."

I couldn't help but get a sickening feeling. The CIA sure as hell had a bad name with the troops. I thought back to college, University of Wisconsin. Those last two years, my fraternity brother, Fred Seibold, and I met with the school's vice president, Luberg, who was also a CIA recruiter. Fred and I both studied Russian, and I had gone on to Middlebury School of Languages for the advance summer course. With our Russian background, Fred and I felt it was the way to go and Luberg reinforced our decision by suggesting we go through the Naval program and their intelligence division, which would also meet our military obligation. He said the Marine Corps had the same dual arrangement. I returned from Middlebury and after graduation a few months later enlisted in the Marines.

Then came the military snafu: my recruiting Captain failed to notify the draft board I had enlisted in the Corps. Because it was a month away, they hadn't written up my active papers for Quantico, Officers Candidate School, where I would be attached to Military Intelligence, Russian division of the CIA.

Because I was drafted, none of this happened. After passing the army physical and being sworn into the army, they turned me around and the Marine Captain re-swore me into the Marines and active duty and Parris Island boot camp was the only temporary solution. It took the intelligence group a little over a year-and-a-half to find out where I was—Cherry Point, North Carolina to be exact. This incident did little to convince me about the efficiency of our intelligence and the military bureaucracy. When they finally found me, in spite of my letters to them, they invited me to rejoin once more for a three-year hitch. I politely declined. I was scheduled to get out in less than six months.

I shuddered when I realized how close I came to being part of an organization that had the likes of this guy in civvies.

I took Jim's suggestion. At 3,000 feet, I gradually increased the distance from Turk's Island as I continued to climb at an altitude that kept me in gliding distance to land in case of engine failure. At 10,000 feet, I plotted my course along the Bahamian Islands to Nassau, Bahamas, our next fuel stop. The rest of the trip was uneventful.

Several months later the news came out with the full story— we had inadvertently dropped in on the U-2 overflights of Cuba in the middle of the Russian missile crises. The U2 was stationed on the remote Turks Island.

U2 Aircraft (photo courtesy Wikipedia & US AirForce)

3. HURRICANES AND FRENCH CUSTOMS

The stories and events of my trips back to South America circulated among our Elm Grove friends. At one social gathering a fellow came up to me and raised his glass. "To a fellow pilot," as he clinked my glass. "Pete Hammond, live in Waukesha. We met before but just in passing. Your wife knows my wife, Pat, sorority sisters or something."

"Hi, I'm Don. You fly? What kind of plane?"

"Beech Bonanza, a good plane to fly down the islands. Always wanted to make the trip. I have the plane and the time. You have the experience of many trips. Would you like to team up? I'd like to see the back country of South America."

And so the seed was sown for the next diamond-buying trip. Jan came up with the idea that since it was a four-seater plane, she would gladly come along to do the navigating, which she was good at.

"Your parents said they would baby sit the kids at the lake, and I would really like to see Nellie, Caesar, the McTurks, and Dutch again. Who knows, it may be the last time?" Jan said. The trip became a threesome.

As did most of the trips, this one started out uneventful with good weather. There was a possible storm brewing up around Antigua and Guadeloupe that might turn into a hurricane—with an ETA at San Juan, about the same time we were due there. That was over a thousand miles out and much could change by the time we flew into that area, so it was not a big concern.

Flying at 10,000 feet, forty-five minutes west of San Juan, and just leaving the north coast of Dominican Republic, I pointed out the black ominous clouds out in the distance.

"They're a long way out. We should land at San Juan before it hits there. Funny we didn't get an update on it out of Ft. Lauderdale. There nothing indicating the size of this one."

The wind picked up, we bounced around considerably, but we accomplished a reasonable landing without too many bumps. So far, Pete flew all the way in spite of the agreement that we would share the flying. I needed the experience to feel comfortable for the short field landings required in the Guiana interior. We were still less than halfway with plenty of flying time left. I decided to hold any comments.

By the time we settled in the hotel, the storm hit full force, and it was pitch dark an hour earlier than normal. This was a major storm.

"Jan, before we go out for dinner, I'm going to call the weather bureau to see just how much of a problem we have here."

I hung up the phone. "It's been upgraded to a full hurricane. They recommend we add some extra tie-downs to the plane. Winds are well over a hundred miles an hour. I'll call Pete and tell him we have to go out in this downpour and secure the plane with more tie-downs—as an added precaution. We want to make sure we have a plane in the morning."

"I'll go with you," Jan quickly added. "We'll wear our rain jackets, but we're still going to get soaked. Three sets of hands will be faster. I've done this a hundred times and I'm not sure Pete has ever done it or realizes how serious this is. For a pilot who owns his own plane, I'm not too impressed. When he pre-flighted the plane this morning, I almost choked. I wouldn't have gotten in until you went around inspecting it thoroughly. Took almost fifteen minutes and I thought Pete was going to have a fit, he was so anxious to take off. Is it just me or was that pretty sloppy?"

"No, you're right. Frankly, a lot of pilots don't bother pre-flighting thoroughly, but then again when you realize the long list of pilots we knew who are no longer around, it speaks for itself. The difference is we're here, and they're not."

We met Pete in the lobby, and the hotel people thought we were crazy to go out in the blinding rainstorm. The palm trees were bending over at right angles and all sorts of things were flying by the window. An Eastern Airlines line crew saw our predicament and gave us extra lines and some five-gallon buckets filled with water. The fifty-pound buckets tied below the wing would be a big stabilizing force. There wasn't anything else we could do except wait it out and hope.

We returned to our hotel rooms, stripped off our clothes, toweled off, and dressed in some casual clothes. We were in dire need of a nice dinner with wine after this ordeal. There weren't many taxis but then there weren't many guests venturing out, either. We finally found a cab driver who agreed to take us to the Old San Juan, wait for us, and bring us back. The meal and ambiance were worth it. The storm was still in full strength, forcing us to creep back to the hotel at ten miles an hour. The windshield wipers were going full bore but not enough to see well. Fortunately, there weren't many cars on the road. Those that were traveled at a snail pace.

When we returned to our room and opened the door, we were hit with the full force of the storm and rain. With Jan close behind, I stepped inside, switched on the light, and shut the door, which stopped the wind. The rain was pouring into the room. Jan saw it the same time I did—the eight-by-three foot window had blown out of the wall and the bottom edge of it crushed Jan's pillow.

We stood in awed silence. Finally, I said, "Thank goodness for our dinner in Old San Juan. I guess the old saying still goes: *When it isn't your time, it isn't your time.*" We hugged. The hotel changed our room and apologized profusely.

In the morning we checked the plane to make sure it survived in good shape and undid the tie-downs before we sat down to the Hotel's complimentary breakfast. The salubrious breakfast and the sunshine made the day appear less bleak. After the harried night, we took our time getting started. Nobody fueled during the hurricane and it was several hours waiting our turn at the fuel truck.

We decided to take an easy pace and instead of trying to reach Trinidad we changed our overnight to Antigua, about half the distance. It was an easy flight. Seventy five miles west of Antigua, I had an idea.

"What do you two think about going to Guadeloupe instead of Antigua?" We could enjoy an outstanding meal on the French island and it's only forty-five minutes longer if we make the decision now and take a straight head-on course. It may be a little dark when we arrive but that shouldn't make any difference."

Famous last words. We landed and parked where a guy in a bright red jump suit guided us to a tie-down spot. We exited the plane and were met by two uniformed French officials shouting at us. They were unintelligible. Pete spoke no French, I a little, so Jan who had three years in college became our spokes-lady. She got them slowed down to where she understood the meaning.

She turned to me. "They're mad—angry."

I rolled my eyes. "I didn't need a translation to figure that out. Is it anything insurmountable?"

"Well, it's what you thought: flying at night, single engine over water is against French regulations, but they seem to be blowing it all out of proportion."

"Yeah, that's because this is a small island, not much traffic, they're bored out of their skulls, and they haven't anything else to keep them occupied."

"That's all well and good, Mr. Pilot, but they're threatening us with all sorts of infractions and fines. The amounts seem high, but they're in Francs, so I don't know how that relates."

"Okay, let them rant on for a few minutes."

I stepped forward, closer to the more aggressive one, and said nothing more than, "*Oui, oui, d'accor.*" Rough translation is that I agreed. Eventually he ran out of steam. There was a silence. First I apologized for my inadequate French and then for breaking the regulations of flying at night with a single engine. They didn't know what and where this was going, which is what I wanted. I raised my finger as if to say something, but left a long pregnant pause instead. I had their full attention. "You are Frenchmen, no?

"Mais oui," they both answered in unison.

"We had a very hard flight and overnight in San Juan when the hurricane passed last night. Then this easy flight…from San Juan to Antigua. Tell me, gentlemen, if you had a choice to land on an English Island and eat English food or go to a French Island and enjoy real French cooking and wine, which would you choose?" I paused and kept my finger in the air.

They both spoke at once. The harsh accusatory voices were gone, replaced by one trying to outdo the other. The customs man was more dominant and pointed to a car alongside the airport. "He's my cousin, Francois, a taxi driver. He can take you to town. I can make a recommendation of the finest restaurant in town. It's not expensive, but it is the very best food. Francois will take you to the hotel, make reservations at my Uncle's, take you there, wait for you and bring you back to your hotel. And if you want, he will bring you back here tomorrow to continue your trip. You, monsieurs and madam, are in for the treat of your life. Bon appetite."

In the exuberance and enthusiasm, he neglected to have the customs and immigration forms filled out. We did that in the morning before we departed.

When Francois pulled up to the restaurant, I thought he'd made a mistake. It looked like a deserted warehouse. He directed us to a side door. We entered and were pleasantly surprised to find a most quaint French restaurant: red and white checked table cloths with a candle on each, the walls stacked with French wines as if we were in a wine cellar. Black-coated waiters bustled back and forth with fold-over napkins on their arms. On the far side a dimly-lit bar was lined with local Frenchmen and women in various manner of dress. It was a scene straight out of Hemingway.

I volunteered to order, but Jan took over. I was dying for French escargots and pointed it out to her on the menu. There was long discussion as to how much we were ordering, but I decided my input was not needed and instead directed my attention to the more serious subject of savoring the full-bodied Bordeaux wine. When the first course came, there were three huge platters of escargots. I was confused. Pete never tried them and wasn't about to. Jan had tried them once and wanted no part of them. The waiter put one platter in front of me and the other two on the table alongside. As soon as I finished one, the waiter picked it up and replaced it with a full one. I ate all three—the first time in my life that I had my fill of escargots. It was heavenly and of course so was the rest of the meal.

If it were only the hurricane and the near confrontation with customs, this would have been a fairly typical trip and we would have continued taking flights with other U.S. pilots, but the next incident in the interior of Guyana sealed the decision to never do it again.

Jan and I discussed Pete's unwillingness to let me do some of the flying, which was part of the original plan. We arrived in Atkinson Field in Guiana, without my once having touched the controls. He simply stonewalled and remained silent each time I brought it up.

"Pete, I'm concerned because you wanted to see the interior. Our next leg will be into the bush strips. You wanted me to do those landings but there is no way without a feel on how the plane handles. The only way I can get that is to do some landings."

"Yeah, yeah, we'll do some landings," and then he shut up and didn't discuss it any further. This did not bode well.

Our first stop was Good Hope where Caesar had accumulated a parcel of diamonds for us. The 4,000-foot Good Hope airstrip was long enough so even Pete could handle it.

Both Nellie and Caesar were there to greet us, and it was a grand reunion. Jan and Nellie had not seen each other in four years.

"Jan, first you have to stay overnight. Then tomorrow we should go to Karanambo. The McTurks insist you come for lunch, and it is especially important, because Dutchie will be there with some diamonds he bought at Marquis *und some utter damn business* he has to settle with Don. Apparently he sold everything in the trade store and has the diamonds for it. He's really proud of that."

Late the next morning, Nellie, Jan, Pete, and I took off in the Bonanza for Karanambo, a short fifteen-minute flight. The ranch was along the river, but the DC-3 airstrip was on higher ground three miles away. We had to first buzz the ranch house, then hope that Tiny McTurk was there and his jeep was working, so he could come fetch us. If everything was coordinated, it would be about a half-hour wait at the airstrip. If not, it would be an hour's walk down the bush trail across the creek to the ranch.

We circled the house. Connie and Dutch waved at us, and Dutch pointed to the wind sock, motioned in a circle and pointed with both hands down the airstrip along the house. Dutch built the airstrip years ago so that we could drop in at McTurks for tea without having them pick us up on the DC-3 strip.

Pete saw the airstrip and nodded to me as if to turn the controls over, and I would land there. I nodded negatively.

"Not without having done one landing in this plane. It's a marginal strip for a Bonanza and the reason I pushed you all along to do a few landings. Let's just go back to the DC-3 strip. They'll pick us up there." Pete didn't like that answer. He flew over once more.

"I can do that airstrip. It's not too short. Done a lot of them like that. No problem."

"Pete, I haven't seen you do or practice one short-field landing in the whole week. This is no time to practice, just go to the big strip."

He wasn't to be dissuaded.

"This is my plane. I know it well, been flying it five years. That airstrip's a piece of cake. I'm going in."

It appeared we didn't have a choice. I was going to monitor this minutely all the way down. He came in on a long final but since he couldn't see the strip from far out, he took my directions for lining us up. He had half flaps on.

"Pete, you know, you will have to have to slow this thing down to below sixty and use full flaps and three-quarter power, right?" No answer. "Pete, are you listening? You won't stop in time unless you slow up and use full flaps. You're still too high."

"I got it, I got it." He hit the flap button again but not all the way. We were too high and too fast and the runway was coming up quickly. He hit the landing gear, a little late I thought, and it slowed us down but not enough to my satisfaction.

"Pete," I shouted this time, "do a go-around and do us all a favor and stop trying to prove something. I don't like what I'm seeing on this approach. Do a go-around," I repeated. He wasn't set up for a proper landing. He should have had the power full off and he still had power, the flaps were not down all the way, the prop should have been shoved in to low pitch, high RPM, for an emergency go around. It was still in cruise. Airspeed indicated seventy, ten knots too high, and we should have been coming in almost at ground level to touch down right at the beginning of the clearing. We were coming in close and still were 50 feet too high.

"Pete!" and I really shouted this time. He was frozen and not doing anything.

I pushed in the prop control for full RPM, while I shoved his hand on the throttle to full manifold pressure, and almost simultaneously tapped the flap lever into take off position, while I hit the landing gear into up position. There wasn't time to release and swing the control column to my side. I roughly grabbed Pete by the chin, shook him and said, "Get off—You're out." And I tore his hand off the wheel and took over. Fortunately, he was still in his catatonic state and didn't fight me.

The RPM went to the red line, the manifold showed full power, the wheels were coming up and the flaps were coming into take-off position as the strip raced below us. In the last few seconds our airspeed had dropped to sixty, near stall, but we were beyond the beginning touch-down point. We were in the marginal disaster zone, overshooting the runway, but not prepped enough for a go-around. I kept the Bonanza a few feet off the runway until our stall horn stopped blaring. When the airspeed indicated sixty, I pulled back on the wheel. I didn't like what I saw. Our flight path was below the tree line ahead.

As I eased back again on the wheel, the stall horn went off again, not adding to any peace and tranquility at this stage. It was going to be nip and tuck to clear the trees. Pete instinctively put his hand back on the wheel, and I could feel the back pressure. Pete wanted to climb— but any more angle at this speed would stall the plane…and it would be all over. I wanted him off and I clubbed his hand off the wheel.

"We're at stall speed, dammit! Keep your hands off and pray!" I shouted and hoped the plane was clean with flaps and wheels retracted, because we were definitely going through the tops of the trees. The landing gear wheels catching on branches would be of no help. The green light on the dash indicated the landing gear was up and I hoped it was correct. The plane handled sloppily, a sure sign we were pushing the stall margin. Just a few more seconds. We weren't high enough. The few extra seconds in the cleaned up plane with full power gave us a few more knots. I had to chance it. I pulled back on the wheel. We came up a few more feet and went through the top branches with the prop clanging as it lopped off twigs, the stall horn blaring, and the thud and scraping of the smaller branches as they were hit by the fuselage. We were through and still airborne. I shoved the nose down a bit to pick up speed and shut up that nerve-wracking stall horn. I made a gradual turn toward the big airstrip and nudged my flying ace.

"You take over and land over there on the Karanambo DC-3 strip. It's 6.000 feet long so you should be able to handle it." We landed. Jan got out and walked over to our pilot.

"Pete, that was inexcusable. I'm not going to fly with you anymore," and she walked off to be alone with Nellie. It was a harrowing experience.

We cut short any stops other than the big DC-3 strips, which considerably limited my buying. But none of us were in any mood to have a repeat of Pete's irresponsible behavior. Jan stayed at Good Hope until the time to leave and then we flew back to Georgetown. The strained atmosphere didn't get much better for the whole of the return trip. Jan and I decided that was the last time we would fly down with someone else in their plane.

4. THE ENCHANTRESS DIAMOND, CALIFORNIA TRIP #1

One of the highlights in working with my brother Bob was the 10-carat marquise blue diamond that Bob bought from Chooch Kaufman, a world-wide dealer. It allegedly was owned by the Countess of Austria. The diamond was of such magnificence as to easily have been part of the *crown jewels.* It was one-and-a-half inches long, beautifully proportioned, grey-blue color in daylight, and an absolute sparkling knockout in the evening lights—a gem that you couldn't take your eyes off of once you saw it.

It was the biggest diamond in Bob's inventory and if it didn't move, his cash flow would suffer. It wasn't a Milwaukee gem, more appropriately New York, Hollywood or Beverly Hills. A bit naïve, off we went with our marquise-shaped *Enchantress.* We stayed in the Beverly Hills Hotel, the heart of stardom, but we quickly learned it was not easy contacting the stars who could afford this rare bauble

We went through several agents without much luck, until we made the decision to go to the agent's office, tell our story in person, show the *Enchantress* to the receptionist, and relate the crown jewel's history. We slipped the magnificent ring on the receptionist's finger. We had her undivided attention. It must be a female ESP thing— within seconds everyone in the office was there to admire or try it on. The result was an appointment to meet Zsa Zsa Gabor, the agency's most likely client. We returned in the afternoon to meet Zsa Zsa in their office. She was more beautiful in real life than in her movies, and her charming, delightful accent captivated my brother and me. She sported a 10-carat round but graciously conceded our blue was a most unusual diamond. It looked great on her hand. But she sighed and said she was not a candidate—at that particular time she didn't have a suitor who would buy this bauble for her. However, her neighbor friend, a Mrs. Doheny, who also loved jewels, might be a candidate. She called and made an appointment for us to visit.

"Darling, I have these two delightful young men here showing me the most fabulous rock. You really must see it. It's one-of-a-kind…even if you don't buy it you'll enjoy seeing this crown jewel." She handed me the phone and I wrote down the directions to Doheny's home.

The agency people briefed us on the Doheny family history before we left. But I was surprised Mrs. Doheny brought up the subject—the family scandal in the 20's of her grandfather's involvement in the Teapot Dome Scandal with Sinclair Oil Company—as if this kind of thing happens in all families. Her grandfather, she said, was the victim of a sensational press and political vendetta of the time. We nodded dutifully, but it was kind of hard to make comments on that subject without sticking one's foot in one's mouth. Silence seemed to be the best solution.

The home was fabulous, large, tastefully and luxuriously decorated. In the back of the house was a sprawling building, her kennel, she explained. She bred and raised pedigree dogs, little white fluffy ones that had an equally fluffy name I'd not heard of before. They ran around the yard yipping when they detected our presence. Mrs. Doheny was wearing a 10-carat emerald-cut diamond. The thought crossed my mind that the minimum allowable size in Beverly Hills was 10 carats. Hers was beautiful but when I opened the box and presented the *Enchantress,* there was an audible gasp from the seemingly unflappable, Mrs. Doheny.

"It's more than I expected…even after Zsa Zsa and you described it over the phone." She walked around the room admiring the dazzling sparkles under all the different light conditions, even going into the windowless library to see the effect of the ring under incandescent reading lamps. It just got better. Bob and I had our hopes up, but alas, desires are not always sales. In spite of the "oohs" and "ahs," we drove back in the taxi without a sale.

It was four o'clock tea time, or more accurately, drink time, in the popular Beverly Hilton bar. In between each U-shaped booth was a strategically placed phone, which suggested that it was important to keep in touch if you lived in Beverly Hills. Every few minutes there would be calls interrupting conversation. Those calls were mostly for the well-known movie stars, part of their PR or *persona* perhaps. Van Johnson was called several times and could be seen strutting past the bar to some unknown destination. Judy Garland was also a regular of the mobile hotel crowd.

The previous day we had an early *tea* and were discussing war games and what to do next. Sitting alone in the booth next to us was a quiet young man who didn't fit the Hollywood profile. *Johnny* didn't have anyone joining him and appreciated the camaraderie, the stories of South America, and was duly impressed with our crown jewel but surprised we were carrying it without a body guard. We invited him to join us the next day.

When we returned from Mrs. Doheny's, Johnny was already there. He appeared more nervous, and we sweet-talked him into telling his story. He said he would be the Master of Ceremonies at an evening talk show the following night, an on-the-job tryout for Arthur Godfrey's show. Arthur was tooling around Central America in his private DC-3 and would be gone for a week. So we talked about the fun and opportunities it presented if he stayed on as the new MC. We parted, exchanged business cards and wished him well.

Several years later, Jan, the family, and I flew into Florida on a short trip and the first stop was McDonalds where we could get a meal for the whole family for under $5.00. Second priority after checking into a hotel was turning on the TV, something unknown in the West Indies.

On the first station we tuned in we saw a guy named Ed McMahon calling out. "And heeeeere's_Johnny."

Donald Haack

"Jan, check this out. Remember the guy I told you about…the one Bob and I met in Beverly Hills who was standing in for Arthur Godfrey? He's the Johnny they're introducing—Johnny Carson, and he's got his own show, now. Good for him… and let's hope he does well."

5. "THE ENCHANTRESS"—TRIP #2

E ven the best crown jewels still have to have a proper home for someone who can appreciate and afford them. It was finding that right buyer. Bob and I didn't have any luck on our first Hollywood trip, trying to sell the blue marquis *Enchantress* to the movie stars.

Jan and I were taking our family to San Francisco to visit her parents. Jan's father, Bill Mills, was the business manager of the San Francisco Examiner and didn't expect to be there too long. The Hearst Corporation needed someone at their Baltimore paper and had asked Bill Mills to go. This would be a unique opportunity for our family to visit Jan's parents in San Francisco and take the Western Pacific Railroad's famous *Feather River Canyon* trip with glass-domed cars.

My friend, C. V. Wood, Tom Slick's (the one I rescued from his plane crash in Wai Wai Country described in *Bush Pilot in Diamond Country*) right-hand man, owned a home in Beverly Hills and invited us to his New Year's Eve bash. That was also the first year in decades that the University of Wisconsin's football team would play in the Rose Bowl game. Jan's father promised us tickets if we came. We couldn't say no.

When Woody invited us for New Year's Eve, he commented about the possibility of becoming engaged to his friend, Edie. Did I have any suggestions? Jan and I decided to take the *Enchantress,* an extravagance that Woody would love and could well afford. At first, Jan was nervous about taking it, but once decided, the bauble never left her hand. She stayed close to her security blanket— me.

San Francisco was everything we hoped for. Bill and Helen took us out to a fabulous dinner in Sausalito, across the Golden Gate Bridge, one night, then to my favorite author, John Steinbeck's old haunts around Fisherman's Wharf in Cannery Row.

Jan and I had no premonition that a few years later we would have dinner with the Steinbecks and that dinner would later change the direction of our lives.

Jan's father took me to the Hearst office where I met William Randolph Hearst's son who listened to Bill explain where his daughter had lived in South America among the savages. It was the first and only time I heard him say anything positive about our living in the jungle in that *damnedmudhut.* Come to think about it, *that* was the second historic event—all in the same day.

Bill had persuaded Randolph Hearst to procure two precious tickets to the Rosebowl and a hotel room in Los Angeles. He did and we arrived in LA in time to clean up and arrange a taxi to Woody's house in Beverly Hills. It was dark when we arrived and the party was well on its way. The front door was open. We rang the bell and walked in. Several people greeted us while we were handed exotic drinks and made to feel welcome.

"I wonder if Woody moved. I seem to recall there was a walk over a stream when you entered his house and there were several pools inside. Maybe I'm mistaking it for Tom Slick's home when Bob and I were there a couple of years ago. Let's go find our host." After several inquiries, a beautiful young brunette came up to us, extended her hand, and welcomed us to the party. I introduced Jan and said we were looking for my friend and host, Woody.

"I'm your host—actually, hostess. Some of my friends do call me Woody, Natalie Wood."

I don't know many movie people but the name was familiar. I must have appeared a bit perplexed. Is C. V. Wood here, too? Are we at the right party?"

She was totally non-plussed. "Oh, C. V. Wood, yes he lives a block away and you aren't the first to be confused. Some of his friends arrive here and some of mine wind up at his house. I hear he's a great guy, but I never met him. Feel free to stay for a while and get to know *my* friends. If you feel you have to rush, I'll have someone drive you over. It's a block, but that's too far to walk. In the meantime, enjoy your drink and happy new year."

Natalie was right; it was too far to walk, and I was glad we accepted the lift. As we stepped over the bridge, I noticed Woody had white flowers floating in the slow-moving water below, and we were greeted like long lost friends by Sharon, whom I met on a previous trip. Sharon was Woody's *Gal Friday* who planned this party.

We were barely across the bridge after a few more introductions when Sharon gasped, "My God, Jan, can I see your ring? I've never seen anything like it, and I've been in Beverly Hills for twelve years." I explained why I brought it and Sharon said maybe it was just as well that Edie wasn't here tonight. "She's still in London. Jan, would you mind if I showed it around, especially to tease Woody with it? Believe me, I will make sure he gets this for Edie."

Jan was relieved not to have to be responsible for the diamond and felt as I did, that the very outgoing Sharon would have everyone in the party ga-ga over the crown jewel. And that's exactly what happened. The party was fabulous. Some of the guests were a bit strange, but many were of the Hollywood group and considered normal in that circle. Woody was the ultimate host, and it was fun to see him again. He made us feel part of the group and insisted we regale them with our tales of South America, bush flying, and Jan's "Annie Oakley" shooting skills. Sometime in the wee hours, Jan had the diamond back on her finger, and we returned to our hotel, thankful for a short sleep before we attended the Rose Bowl game later in the day.

If we had to pick a Rose Bowl game, we couldn't have chosen a more exciting one—specifically the last seven minutes, as one of the most memorable games in Rose Bowl history. In the third quarter UCLA was trouncing Wisconsin 44 to 7, not a happy situation for us U of W graduates. Then with seven minutes left, the Wisconsin coach gave the new quarterback, Ron Vanderkelen, a free hand and in the last few minutes, Wisconsin scored five times coming to within seven points of UCLA. Vanderkelen was unbelievable. Nothing stopped him. Every pass was a winner. Even when tackled, he would throw or flip the ball to a receiver. He ran, he was tackled sideways, upside down, and he still found his receivers. After scoring, Wisconsin onside kicked, regained the ball, and repeated the performance. I had never seen anything like it in football and apparently neither had anyone else. It truly was a game of a lifetime. As usual in the last ten minutes of a lopsided scoring game, many fans were bored and exited early—to miss the game of the century. Wisconsin didn't win, but I could never be more proud of a team in its effort on that day. Fabulous.

In each of my trips to California, and there were four, I would never have a dull trip and this was no exception: The Wisconsin vs. UCLA all-time classic game, meeting Randolph Hearst, Natalie Wood, and showing the *Enchantress,* which Woody eventually presented to Edie Hollingsworth as an engagement ring.

6. DIAMOND SAFARI IS BORN

Our neighbor, John Warner, listened intently as we described our last plane trip to South America. "And you're still getting requests from other friends to join you on the *next trip*? Mind if I sum up what I'm seeing here? The partnership you came to Wisconsin for, Diamond Designs, has not turned out to be a partnership for you after all. The diamond trips to South America are benefiting a business you're not a partner in. You're running two months behind on your salary. You're having requests from people who want to join you on your expeditions. Does that just about sum it up?"

I had to laugh. "Yeah, sounds pretty screwy, doesn't it? And maybe a little grim, too."

"There may be an interesting opportunity here. I have an idea. Let me get back to you in a few days." John was in advertising and had more creative ideas than anyone I've known. They flowed on any subject brought up. This was no exception.

I quit the job/partnership with my brother. It was not one of our greater moments: three months with no salary, Jan was expecting our fourth child, and the bank account was perilously low. In accounting terms, it was empty.

"Jan, there's an old saying, *When things look the darkest, put on a good face*. The Oschmanns from Conover offered us their cabin several times, and when Herb and Maggie came in last week, they repeated their offer again. We've got fifty dollars cash in our piggy bank. Let's go up North for a few days. The leaves should be changing—the best time of year to be up there."

Jan's first reaction was expressed in her face—I had lost my marbles. But it quickly took on a positive reaction. "Maybe you're right. Things couldn't look much worse right now, and a change of scenery might just be what we need. Where is their cabin?"

"On a lake a few miles out of Eagle River." I made a phone call. Herb, the perfect host, explained where the key was and reminded me that we were going up just in time for the fall color change. The next morning we packed the Mercedes in record time and were on the road soon after breakfast. Our mood was infectious, and Diana, Tom, and Todd were caught up in the excitement of going into the north woods and being in a cabin right on the lake.

It was a five-hour trip. Jan and I took turns driving. I had the last two hours. The leaves had started to turn, making the trip one ooh and ah after another, admiring Mother Nature's artistic touch. I was driving alongside a lake on the right, and the road was lined with tall colorful trees on both sides. Relaxed, I was caught in the reverie of admiring the colorful and peaceful scenery, when I noticed a car behind us that I hadn't seen before. I glanced up in the mirror, wondering where the vehicle came from so quickly. I hadn't seen side roads after I approached the lake. The road made a gradual turn following the contour of the lake. The car, driven by a woman, was a white station wagon with a sign on top. Jan noticed my glancing up at the mirror.

"Is anything wrong? You're not going over the speed limit. Is it a police car behind us?"

"No, just kind of strange. That car just came up on me so fast that I hadn't noticed it."

Jan's curiosity got the best of her and she turned around in her seat. "What car?" she asked. "There's nothing there."

I checked the rear view mirror again. No car. I didn't see any roads where it could have turned off. That bothered me. Was I day dreaming a little bit too much? Either way, the unsettling feeling lingered for quite some time until we reached Eagle River town limits.

Jan was observing the Lion's Club notices. "We sure came up here at the right time. The Lion's Club is having a *Colorama* event. There are people and signs all over town. For the price of their picnic dinners we might as well come back here and eat. It's cheaper than we could cook up, and we won't have to fuss with any meals tonight. Let's go find our cabin, unpack, and come here for the festivities."

I agreed. Ten minutes later we pulled off the main road, drove a couple of miles on dirt, and spotted two deer ambling slowly off the road as we pulled into the Oschmann's driveway. We arrived on a secluded lake with a picture-book log cabin at the shore's edge. We unpacked and inspected the rustic cabin, which was bigger than anticipated—the kind you would dream of in a secluded northern woods setting. It couldn't be better.

We checked out the lake and found the canoe our friends told us to use. That would be the first thing in the morning. Everyone was hungry, so off to the Lion's Club Colorama. A few minutes later, we were in line filling up our plates and being welcomed by the local Lions as if we were long lost brothers. It was a great way to start our retreat.

"Jan, I need five dollars. I'm going to buy a raffle ticket."

"You have to be kidding. We're short on cash and you're going to spend five dollars on a raffle ticket? I see they're selling tickets and raffling fifty prizes, but we don't need prizes."

This went on back and forth. I was trying not to give the real reason—it wouldn't sound logical. Finally, I blurted out, "I know this sounds crazy to you, but remember the car I was telling you about when we were going around the lake? Check out the first prize…it's that car, the white station wagon with the sign on it."

"Yes, so what?"

"We're going to win it, and you'll be driving it back."

"That's the silliest thing I've ever heard. Do you actually know what you're saying? I can't believe it."

I was not backing down. I bought ticket number 499. I also had a very distraught wife. I pointed out that they sold less than 500 tickets, a fact that the Lion's Club also understood. In hopes of selling more, they extended the drawing hour.

It was getting late in the afternoon and there was a chill in the air. The raffle finally started and the man with the ticket drum called over and asked if our daughter would draw the tickets. For a nine-year-old that was quite an honor, and she hopped up on the stage. Tom and Todd were playing on the teeter-totter. Tom slipped and cut his lip on the bar. He was crying, Todd was cold. Jan was more than ready to get back to our cabin and build a cozy fire.

"They're getting down to the last three prizes. We have to get Diana back here."

"Why? As a matter of fact that's a good idea . . . and go back to our cabin. I've had enough."

"Diana can't draw any more tickets." We're going to win, and we can't have her doing the drawing. Just hang in there a few more minutes."

"No, let's go now. Tom is still crying, Todd is cold, and Diana is finished drawing tickets. We don't have to be here to win the car, if that's your crazy theory. They can notify us and we'll be back in the morning."

"No, it doesn't work that way. We have to be here for the drawing. Okay, you go ahead. I'll stay here and either you can come back for me in an hour or I'll find some way to get back to the cabin. But I'm not leaving. We're going to win the car."

I noticed that Diana and Tom were taking second glances at me. The third and second-place prizes were drawn and the grand prize finally came up. The man on the platform spun the wheel a couple of extra times to thoroughly mix up the tickets. Everyone was huddled in close, hoping their ticket was the lucky one. I edged my way through the crowd until I was directly below the drawing wheel.

The wheel stopped, the second man reached in and handed the raffle man the ticket. "The winner of the grand prize, this brand new

1964 Ford Station wagon, is number 4-9-9!" and he held up the ticket. I held up my ticket and handed it to him. He was surprised the winner was so close. There were audible groans from all the other ticket holders. "Congratulations. This car is yours. I'll bet you're the most surprised person in this whole Colorama country?"

I started to say, "No, I knew all along I was going to win," but thought better of it. No one would understand that kind of reaction. Instead, I heard the words come out, "Yeah, I'm the most surprised person in this county. Thank you and for all the good things you Lions do. This was quite an occasion. Thank you again." I gave my name and address, and we agreed to do all the paperwork in the morning. I walked back to my family who were in a state of disbelief and euphoria.

"You really knew all along, didn't you?" Jan asked. "Let's go build a fire and celebrate."

Four days later, I was again driving the Mercedes. Diana was in the front seat with me. Tom and Todd were asleep in the station wagon. We came to the same lake and as we rounded it, I glanced up into the rearview mirror. There was a lady driving the white car. I recognized who it was—Jan. The Lion's Club sign was still on the roof. This time the image did not go away. It followed me all the way to Elm Grove.

I couldn't help but think that this was providential, a sign that our fortunes were changing for the better and something was going to present itself to us.

The night we returned, John Warner rushed into our living room and pulled out his briefcase. I could tell he was in high gear with some not-to-be-held-back ideas. He spread them out.

"A diamond safari! Adventure, travel, exploration, finding one's own diamond in the middle of South America. What more could you ask for? I've a mock-up brochure with some marketing ideas—an advertiser's dream." He spread it all out across the table. "I priced it so high, it will discourage your friends from wanting to join you. If by some chance you still get takers at that exorbitant price, well, you have a going business."

To all of our surprises, we received inquiries, *takers* and lots of them.

And that's how *Diamond Safari* was born. John was right about an advertiser's marketing dream. It began with feature articles in the *Milwaukee Journal* and talk shows on three of our local stations. Two Chicago stations requested interviews for their talk shows, which resulted in Abercrombie and Fitch Travel Tours giving us a center-fold feature in their travel catalogue. That was followed by a five-page feature in the Sunday Magazine section of the Chicago Tribune. We were off and running on a new career whether we wanted to or not. In the midst of all this, our fourth child, Julie, arrived in the Milwaukee hospital, the same one I was born in thirty-four years earlier. Our fortunes indeed change more than we could have expected. Jan and Julie were doing well. In between baby-caring, Jan managed to handle the inquiries, talks, and bookings of our new enterprise.

Within four months of the sit-down-at-the-table safari idea, we were up and running our first expedition. I purchased a four passenger Cessna 180 and then worked with a couple of people familiar with the new product called *fiberglass*. One of my first prospective customers was Frank Hawk, an engineer with enough money for the safari, but who wanted to stay on a permanent basis. He helped build the pontoons and offered to accompany me on the Cessna flight to South America. Being an engineer-mechanic, Frank could keep the aircraft maintained and any machinery running. He boasted he was a first class cook. For him, it was not just the two-week safari. He wanted to stay permanently. I took him on, tentatively. He stayed three years.

Together we designed and built four-foot sections of the dredge pontoons that could be stacked and flown in the Cessna, and then be quickly bolted together. These would be the flotation pontoons for the diamond dredges. In the past, the biggest difficulty in diamond dredging was the inability to easily transport barges that carried the suction pumps and diamond-separating equipment. Large trees had to be cut and hollowed out by hand for flotation pontoons.

For the divers, I ordered the modern hookah diving equipment that was safer and more maneuverable than the existing deep-sea, hard hat diving equipment.

On one of my previous buying trips, I met Glen Rhine, an engineer in Guyana who was fabricating light-weight suction pumps and sluices that were quite effective in separating out diamonds. I placed our first order with Glen, who guaranteed completion before I arrived. The combination light-weight dredge and sluice, our portable quick-assembly pontoons, and the modern diving equipment, along with the airplane for transportation, would make for much more efficient diamond prospecting in Guiana.

As we finished stacking the fiberglass sections and crating them for shipment, we received another call from a prospective client, Jim Patlow. He explained he didn't have the funds for the safari, but had an amphibious (land and water) airplane, a Seabee. He was a pilot and a qualified aircraft mechanic. He would like to join us and wanted to know if we could work something out to include him on a permanent basis. After several discussions with Frank, we checked Jim's references, which confirmed he was only twenty-four, did indeed have an A & E (Aircraft & Engine) license, and was eager. I talked to his parents, and they came across as first class people. They were concerned about his going to South America, but if Jim went with us, we could look after him.

"He's a hard worker, but he gets a little wild at times." They didn't answer my questions as to what "a little wild" consisted of but assured me it was nothing serious. I would find out what "a little wild" meant later, much to my mortification.

Patlow's plane needed modifications for better performance, a radio and some mechanical work to be airworthy. I arranged and paid for those improvements. We made up a contract. The family lawyers and his parents okayed it. Patlow and I signed and we now had another airplane, pilot, and mechanic. Patlow, who had never been out of the 'States, would fly down the Bahamas and the West Indies islands with us. We had a deadline for the aircraft to be up to standard in a matter of weeks and fortunately the timing worked out well.

Our two-week Diamond Safaris billed out at $1200. To put that in perspective, it was possible to buy a middle to upper-end brand new Chevrolet for $2500. The $1200 fee did not include airfare to British Guiana, but it was all-inclusive once clients landed at Atkinson Field, British Guiana. There were no extra fees.

Two weeks after our arrival in Georgetown our first safari guests were scheduled to arrive, enough time to assure all the amenities were in place. We made arrangements with the government for our departing safari guests to be taken to the Lands & Mines Department, declare their diamonds and pay a $5 export fee. They received documents allowing them to carry the sealed package with them and import the rough diamonds into the States without fee or duty.

We were ready to have our first safari guests.

Winning Car & Family

7. THE FIRST SAFARI

The completed fiberglass hulls, resin, and woven-roving glass cloth needed to build more pontoons were crated and shipped to B.G. to arrive a few days before Frank, Jim, and I flew in by plane. We landed in Florida where we met Jim Patlow and family, examined his Seabee, and completed a thorough check-out flight. Frank and I put extra fuel tanks in the both airplanes and jury-rigged a hose and pump, so we could re-fuel in air to reduce the time-consuming re-fueling stops down the islands. A few days later our two planes were on the way. First destination was Great Inagua, the southwestern end of the Bahamas where we would overnight.

Surprisingly, that trip was one of our nicest and least eventful flights. Four days later we arrived in Georgetown, British Guiana.

Mac Wilshire, the owner of the Woodbine Hotel, was ready for our group of eight tourists who were to arrive in two weeks. They were to overnight in Trinidad at the Belvedere Hotel. The affable East Indian manager, Sonny, was going to see that they had a good night out seeing the Trinidad night spots. It would be a warm tropical setting and a dramatic change from the freezing temperatures of the Midwest.

The interim two weeks were spent checking in the two aircraft with Alex Phillips, the Director of Civil Aviation, and obtaining the necessary permits to fly in the British Guiana interior. Supplies were purchased and camps set up, awaiting the arrival of our first group of prospectors. Frank made up a list of the food and cooking utensils and I purchased everything else. Glen Rhine had completed two 3 inch suction dredges that were very portable and easy to use. I packed those into the Cessna 180, along with the Brazilian hammocks, mosquito nettings, rain canvases for over the hammocks, surukus (diamond sieves), gold pans, first aid kits, suntan lotion, halogen pills for purifying drinking water, gasoline, oil, and tools. And of course, diamond tweezers, 10 power magnifying loupes, and portable diamond scales. I flew in three Amerindian guides to set up the camps for the tourists.

A cooking tent, dining tent with Frank's cooking utensils, folding canvas captain's chairs and folding aluminum tables were set up in one of the camps, which was located above the spectacular Kaieteur Falls, the biggest in the country. The second camp was the Jungerson's adobe and grass house in the Echilebar area where Dutch and I had prospected years earlier. It was good for both diamonds and gold and the water was shallow and easy to dredge. The first two safaris consisted of eight guests. The first week they stayed at the Kaieteur camp and the second week in Echilebar. Later when we had sixteen guests, they were rotated between the two camps. Three days for R & R were scheduled in the Rupununi at either Gorinski's Good Hope Ranch or at Pirara, the Hart's spread further south.

Wheating & Richter Company found us a small portable kerosene fridge that we carried in and placed in Frank's kitchen tent. That would provide several ice cubes each for our drinks before dinner. We didn't want too many hardships. There was very little duty on liquor, a case of Dewars fine Scotch whiskey cost $20 U.S. The local Demerara Rum ranked among the finest in the world and cost .75 cents a bottle. It would not be a dry safari.

Two weeks later, the first safari prospectors arrived. I scheduled a leisurely tour around Georgetown and its famous Stabroek Open Market. That one leisurely day in town helped them acclimatize to the humidity and temperatures at 5 degrees above the equator. The next morning BG Airways' amphibian Grumman Goose landed five of them on the river above Kaieteur Falls. I transported the remaining three to the landing strip that we'd made along the Potaro River above Kaiteur. That was our first camp. The breathtaking view of the falls was a great beginning—by far the most beautiful and spectacular sight in the country. The flight itself was different in that once we left the low coastal area, the next 150 miles were above rivers and thick jungle, or *bush*, as it was locally called. The terrain changed dramatically from jungle to 4,000 foot high cliffs— the escarpment that Kaiteur Falls plunged over. As we flew above the plateau that extended north and south there was another view to the west—the two-mile high Kukenaam Mountain jutting straight up out of the bush.

On a clear day and further west it was possible to see another equally high mountain, which the Indians called *Awon Tepui*, meaning Devil Mountain. It's better known as Mt. Roraima, the boundary of the three countries: British Guiana, Venezuela, and Brazil. Roraima was made famous by Sir Conan Doyle's 1900 novel written about this plateau. *The Lost World* told of a fictitious civilization living on top, isolated from the rest of the world for thousands of years.

The first small dredge and camp for our prospectors was set up in a tributary creek of the Potaro above Kaiteur Falls. This was a diamond-bearing creek and with help from our Amerindian crew using the new small dredges, our *prospectors* were sure to find diamonds. The second day, the first ¼ ct was plucked out of the diamond sieve, raising the level of work into a surging frenzy to find more.

From experience, I knew that standing over a dredge all day, taking turns diving and standing in shallow water, gets old pretty fast. We kept the pace going. One day was set aside for a flight to Mt. Roraima. Flying between the huge plateaus that disappeared into the clouds made us feel like miniscule bugs in comparison. The majesty of the scenery during those flights was so overwhelming it left everyone speechless. There was complete silence on most of the journeys. Later back at camp, words came pouring out, trying to describe the emotions they'd experienced in this other world.

One other day some of our guests were flown to the Echilebar River near Monkey Mountain where Dutch and two Amerindian boys set up camp. They had a portable dredge in the river that consistently produced diamonds and small gold nuggets. Dutch and the Amerindians stayed in the house that my friend, Jungerson, and his son built. Our safari guests stayed there and in return, I kept Jungerson's caretaker supplied with food and necessities— a typical solution in these isolated bush regions.

Everybody helped everyone else and our lives were intertwined in one way or another. It was the only way to survive in this beautiful but unforgiving land.

Diamonds were there, but the earth was not giving them up easily. We all learned that very quickly.

Three days were spent in the Rupununi with Caesar and Nellie Gorinsky at their Good Hope Ranch, thirty miles north or at the Pirara Ranch where the part-American part-Wapashani Hart brothers lived. The Rupununi ranching country's sunny, flat, arid, open atmosphere was a welcomed relief and contrasted the hot, humid, dark, mountainous jungle. The Rupununi's distinctive Brazilian character of parties or *Fetes*, as they were called were a part of that scenery. One only had to mention the possibility of people getting together and voila—a fete, Brazilian style.

The logistics of moving eight people with two four-seater planes was planned in detail beforehand. For the Rupununi outing, I had everyone rendezvous at the Jungerson's Echilibar house, our jump-off point. With hammock hooks in the walls and army cots and blankets we could spread out at Jungersons. There were no mosquitoes so we didn't need nets. In the first safaris the sleeping arrangements were not a problem. Our guests were young men between the age of twenty-five and forty-five. Later we had several older couples in their fifties and we arranged more privacy for the ladies by adding tents alongside the house.

It was at the Jungerson's that we observed an unexplainable event. We had eleven sober witnesses that included Frank, Dutch, and me. It was seen the first night before our evening cocktail hour, thereby eliminating the theory that we might have been *in our cups,* as the British say.

As our usual ritual, some of us had taken baths in the creek, while others enjoyed our makeshift shower from an overhead gasoline drum. We changed into fresh clothes and were congregating around the table for our evening drinks. Dutch rationed out two ice cubes each from the kerosene fridge. Larry called us from outside. "C'mere, guys. What am I seeing out there?" and he pointed to the dark northern sky. There is no twilight this close to the equator and the sun had just gone down. The landscape was silhouetted in the sky by the faint light of the stars. A light moved across the horizon. "Hey, Haack, you said no planes could fly in this country at night. So what're the lights out there?"

I was out first and in seconds everyone was gathered in a group. The Jungersons had chosen this location, a hill with no trees, because of its great panoramic view.

I watched the light tracking across the sky. "Strange. If it were a plane close to us, we would hear it. In this back country with no distractions, you can hear an engine miles away, even before you can see it. And if it's far away, there is no way it can be a plane. Look how fast it's moving along the horizon…no plane is that fast."

At that moment the light suddenly stopped, changed color, and got bigger and bigger. . . and headed toward us. Then it stopped again, shot straight up, changed into a red hue, and disappeared only to reappear at right angles from the East. We all speculated at what we were seeing. It was big, then changed color and receded into a small dot that shot across the horizon. To be that far away and travel that distance on the horizon would take mind-boggling speed. There was a total absence of sound in the movements, even when it came right at us. We guessed that at its closest it was less than a mile away.

This spectacle lasted over a half hour. At the end, "it" was very large and close. Then, from a large red ball, it receded into a small white dot that disappeared in the northern horizon. We stood in a silent daze. Everyone came forth with a theory. We analyzed each one. None held up. We really didn't know what in the world or outside the world we had just witnessed.

Late into the night we exchanged thoughts and notes about the phenomenon. In the light of day and over breakfast, we decided that with all the notes we had, even if we all signed a composite final draft, it would appear that we were hallucinating from drugs or drink. The decision was made not to send a documentary of the previous night's events to National Geographic unless there was a re-occurrence a second night. Then we could photograph "it" and send the complete documentation. The light display did not occur the next night, but everyone agreed the unusual event was one of the highlights of the two-week safari. We never saw it again.

Donald Haack

Kaieteur Falls, Aerial View

Jungle Scene, Don cutting path. Note khaki clothes are Dark brown (wet) from rain & sweat.

Cessna 180, Harry & Don with supplies. Diamond surukus in foreground.

Frenchman testing sluice

Dutch at Jungerson's adobe & grass home

Tourist suiting up

8. DUTCH, THE MINER

Spending the few days with our safari group at Jungerson's Echilibar house brought back vivid memories of the first time I went prospecting with Dutch. Caesar suggested that we invite Dutch to live with us in Marquis. It would be a good arrangement all around. A couple times a year when the rains became unbearable, Dutch came out of the *mines*—a ten day walk from his favorite prospecting area, to the Rupununi. Lately he wasn't walking too well. If he stayed at our house, he would be five days closer to the mines.

Dutch felt it was his duty to introduce me to his favorite haunts. He had brought it up several times during our cocktail time before dinner.

Jan said, "You might as well go up there with him. He seems to have had the experience and some good diamond finds up in that area. Who knows, with the new lightweight diving equipment, it may be the right time to see what Dutch was doing?"

Jan wasn't aware of it at the time, but she was predicting events that would occur five years later.

"After all, you know Dutch doesn't dive because he can't swim a stroke. Some of the deep spots in the Echilebar may be virgin territory." She chuckled. "And you know Nellie told us, every trip Dutch returned with a good cache of diamonds, in spite of the fact that he lost as many from his shirt pocket when he bent over to wash up in the mornings. Besides, it will be a change of pace for you. You've been going at this project non-stop for the past three weeks. There's a lull now, and it would be a good time to check out *Dutch's Diamond Digs*," as she called them.

Dutch and I took our hammocks and a light back pack that Jan stuffed with home-baked bread, hard cheese, onions, a couple of cans of bully beef, coffee, tea, brown sugar, a small tin of Carnation milk, and some halogen pills to purify water, all of which would keep without refrigeration. Dutch took a bag of farine, sugar, and his plug of tobacco. I carried the surukus, the diamond sieves, and Dutch had the gold pan. We packed everything in the TriPacer and off we went.

Twenty minutes later we landed at Jungerson's airstrip. Dutch showed me where to park the plane, and after screwing in our anchor stakes to tie down the wings in case of a high wind or storm, we were off in the direction he pointed. I checked my compass—we were heading due south.

"Vell, ve saved a day's valking coming here instead of Monkey Mountain. Ve're going to da flatlands ahead. Some goot creeks der." He led. I followed. Periodically, I checked my compass. Our heading was erratic, but I attributed that to the fact that we were not on a trail but were following the small ridges, which made for faster travel. He held up his hand.

"Ve shtop here before ve cross ofer." And he pointed to another ridge a short distance away but separated from us by the thick underbrush nourished by the water below. The knoll we were on was the highest point and provided a panoramic view of the valley Dutch had described earlier. The creeks from this basin drained into the Echilibar River, and he was certain they were the source, the *mother lode* of diamonds and gold. "But furst ve haf refreshments," as he walked over to what looked to me like an apple tree. It was laden with bright red fruit with a brown u-shaped object at the bottom. "Dat's a cashew tree. Pick a fruit, remove da cashew from da bottom and bite into da bright red part—it's sweet. Don't handle da nut on da bottom. Burn your fingers. Gotta roast it furst, den crack it open und pry out da nut. Lotta work. Dat's vy dey so expensive." He was right, the cashew fruit was delicious and thirst quenching. We ate three apiece. I asked how a cashew tree grew here in the middle of nowhere. He shrugged his shoulders and explained it must have been the seeds in bird droppings. Paramakatoi, twenty miles to the North, was the closest Amerindian village where they had cashew trees in abundance.

After a short rest, we clambered down into the dark, thick bush, which Dutch assured me we should cross in a few minutes. If he said five minutes, I calculated it would be twice that. Dutch hadn't worn a watch in thirty years. His sense of time left much to be desired. With no trail to follow, we had to clear our own path through the thick vegetation. We took turns chopping with our machetes while Dutch continued to point out the general direction we were to go. I should have been more observant, but I was caught up in the tales Dutch related about this area. I did, however, take a compass sighting from the other ridge when Dutch pointed out where we were going—95 degrees, an almost direct easterly heading.

The great contrasts of this country always amaze me. A few minutes ago we were in blinding sunshine, hot and dry . . . walking over dried, brown vegetation. Now we're in a little valley, dark and humid with thick, lush vegetation that challenges anyone trying to trespass. Within minutes our khaki shirts and trousers are dark brown from perspiration. The humidity hangs like a wet sponge. Have we been chopping for hours or a fraction of that? I glance at my watch. We're in this dark patch of bush forty-five minutes. Something is wrong. The ridge simply isn't that far away. I take out my compass. The path we're on is a 180 degree heading. In less than fifty yards, Dutch has drifted off course a full 90 degrees. Instead of crossing the gulley, we're going parallel to it, and we could go on for days in this semi-swamp and maybe never come out.

I pointed out our mistaken heading to Dutch and took over the navigation. I couldn't turn east now without knowing how far south we were. We might head into another valley. I back-tracked to where the trail indicated a westerly direction. It was at this point that we should have stayed on the reciprocal, a 90 degree course. We commenced our path-clearing and within a few minutes arrived in the open area on our destination ridge. I often wondered how Dutch survived twenty years in the bush without a sense of direction, timing, or being able to swim. Somebody up there was looking after him, a theory the Indians had subscribed to wholeheartedly. They believed Dutch led a charmed life, and they welcomed him whenever he approached their villages.

We clambered up the ridge into bright shining daylight again. At our position four degrees north of the equator it took several minutes for our eyes to adapt to the extreme brightness of midday sun. We continued south along the ridge as it gradually became less steep and ended in thick bush at the intersection of two creeks. Dutch took off his backpack and sat down.

"Yah, ve're here. Time to rest und have sumpin to eat." He opened his backpack and removed the bags of farine and sugar. "Ve'll shtart prospecting here at dese two creeks." I started to unpack the food Jan had packed, but before I even had it out of the bag, Dutch handed me a tin cup. "Yah, here's a goot lunch." I looked inside the cup—farine, water, and some sugar. Dutch swirled his cup a few times, put it to his lips, tilted his head back and emptied the cup. I took a sip. The liquid had a fermented taste. The farine had the texture of small gravel. I tried to bite down and chew it. Dutch intervened. "No, no, you break your teeth. Joost swallow it down and it vill soften in your stomach." I tried without success, again and again. "Take big gulps und it'll go down, don't vorry." I finally got it down. Dutch got up. "Okay, now ve et, let's get to vork." I was still in a state of digestion shock. I managed to pick up the suruku sieves and follow him to the creek.

We walked into the creek and I was surprised how cold the water was. Dutch explained it was both the altitude and the thick foliage blocking the sun from shining on the water. He carried a small shovel with a sawed off handle that he must have stowed in his backpack. I hadn't seen it before and quickly realized that without it we really couldn't do any prospecting here. We sure couldn't scrape up the gravel in our hands. Good 'ol Dutchie, I thought. He had all the basics. I put the sieves along the bank in the shallow water and Dutch filled a couple shovels full onto the sieves, but not before he placed his gold pan below the sieves. I swirled the sieves around and around and up and down to float the gravel and give any heavier material like diamonds a chance to fall to the bottom and middle of the concave sieves.

I easily screened out the course material and tossed those stones out. Anything in there would be 20 carats or more. Not much chance of that. I worked the second sieve with the same negative results and finally got to the sieve where we would expect to find our diamonds. I did this much more carefully, every couple of turns, scraping off the excess gravel until I had a pretty good concentration. I was surprised and Dutch was disappointed when I picked out a tiny 5 point diamond, $1/20^{th}$ of a carat. Finding one on the first try was exceedingly unusual, but Dutch only shook his head. "Der's more und bigger vuns in here. Dat vun is nutting, you see." Before I could add another shovel of gravel, Dutch stopped me. "Vait, ve haven't checked da gold yet." He bent over and picked up the gold pan from below the bottom sieve. I had forgotten the pan. "How are you mit a gold pan? My back isn't as goot as it used to be."

I reached over and took it from him. "Yeah, I've had some experience with a pan, lemme have it." I swirled it around and removed the larger stones from the outer rim and continued until I was down to the small stones and sand in the bottom. Gradually, I swirled the water inside until I had a two-inch diameter concentration in the center. A few specs of gold would indicate a yield of three to five dollars per yard processed in large quantities. There were more than a few specs. There were tiny pebbles of gold that would yield ten to fifteen dollars per yard. The breaking point of mining in those days was about $4.50 per yard in average mining conditions. Dutch thought the gold alone would pay for any operations here if this were a representative sample, and he claimed it was even better in other areas here.

We spent the rest of the afternoon digging in different locations on the creek with much the same results of diamonds and gold. We stopped well before dark to set up our hammocks, netting, and campfire for dinner—a real meal. I told Dutch I was surprised that I didn't feel hungry working manually with only a half a cup of farine for sustenance. He explained that when the farine was moistened in your stomach, it swelled up and made you feel full. *So much for the gravel lunch.*

In the morning light, over a cup of strong coffee laced with Carnation milk and brown Demerara sugar, we examined our cache of diamonds and gold. We counted over seventy *colors* of gold, the small flakes and sand-like pieces. We had seven diamonds. The largest was 1/10th of a carat. Although small, they were of good quality and gave us an indication of what to expect from this area. It would take small portable equipment to work on a commercially-paying basis.

Five years later, this is where I brought our safari guests. Our 3 inch portable dredge sucked four yards of gravel per hour. With four guests and three Indians to do the final sieving, it was easy for the guests to produce one carat of diamonds each. As the dredge worked deeper into the sand and gravel, the diamonds extracted were bigger, and we were producing ½ and ¾ carat stones for every ten small ones.

Dutch and I spent the next morning prospecting as we did the night before. The morning production, to our satisfaction, was slightly better. After lunch we packed up and returned to the plane.

"Haf you efer been to Awon Paru, da airstrip on the B.G side across from King strip, da settlement in Brazil?" I hadn't, but I'd heard about it, and it seemed like a good time to explore another landing strip along the border. Dutch cached some equipment there on his trip several months earlier, and now with the plane he thought it might be a good opportunity to stop and retrieve it. It could be months before he got back there on foot again. Dutch explained it would be a five-day walk. The Tri-Pacer cut the time to twelve minutes. Our airplane literally became a time machine, shrinking traveling time between villages. There was never a mention of kilometers or miles. Distance was measured in hours or days walking. It wasn't unusual for *six walking days* to be reduced to fifteen minutes of flying.

A few minutes after we landed, two boys from the village arrived, greeted Dutch and acknowledged me in a small way.

The fact that Dutch arrived in an airplane enhanced his status, not mine. I was only the pilot. In broken Portuguese and Macusi, Dutch explained what he needed. They took off for the village, ten minutes from the airstrip. Several more boys showed up to see the airplane and greet us.

I noticed a set of beads hanging on the waist of one of them. On closer inspection, they weren't beads but snail shells. I had never seen a snail in these parts. With many hand motions and some help from Dutch, I finally determined that there were snails in one of these creeks, and they were good to eat. *Escargot, I thought. Haven't had that since Karl Rasche's Restaurant in Milwaukee.*

While we waited for the boys to bring Dutch's equipment, I bargained with them to bring me a sack of snails in exchange for a bag of candy I always kept on board for just this kind of contingency. In less than thirty minutes, we were winging our way back to Marquis, Dutch with his equipment and me with my escargots. I was ecstatic at the thought of having this delicacy. Jan was not impressed. Escargots had never been her thing. Nevertheless, I peeled and finely chopped several cloves of garlic, sautéed the cloves in butter, and poured it over the snails before baking them in our kerosene oven. I made a shaker of Beef Eaters martinis, which Jan did partake of while I feasted on my escargots ala Awon Paru. Sometimes it's the little things that make life sweet.

Donald Haack

Dutch at Good Hope

9. INTERLUDE AND PR

The first safari went off without any hitches, and it would be three weeks before the next. Jan had lined up some publicity during that time, so I took a commercial flight to Wisconsin, left Frank Hawk in charge of equipment and Jim Patlow to do commercial flying with either of the airplanes. Frank would help Jim with the necessary aircraft maintenance. I left feeling everything was in control. Nothing is as it seems.

When I arrived in Milwaukee, Jan was full of questions as to how the first safaris went. Did they have a good time? Anyone injured? Did they all get some diamonds? How did the food and sleeping arrangements go? Eventually I answered everything to her satisfaction. For a start up project in a remote area of a foreign country, it could be considered a roaring success—which was the consensus of the first group.

Jan had scheduled three publicity events: one was a newspaper article, the second an interview on WTMJ where they were having an ex-heavyweight boxer on the *Heini and his Grenadiers show*, a long-running classic program featuring the German band that was a popular radio hit for years in Milwaukee. It was taken off for a couple of years during WWII because of the negative German sentiment. Heini had come back to an exceptionally appreciative audience and Jan thought that being on his program should get good safari exposure. The third was an interview on WGN, one of Chicago's prime talk show stations.

Jan had her hands full with our four children and couldn't get a daytime sitter so I went alone to the WTMJ interview. I was directed into the studio waiting room. Heini was doing his thing with polkas, waltzes, and the German *oom pah pah* he was so well known for. The audience loved it. There was an older man sitting outside the studio listening and watching the band. I figured him to be the heavyweight boxer Jan had mentioned. As I approached, he pointed his finger at the band with an approving nod of his head. During the first break, I explained to him who Heini was and the history of his Grenadier Band in Milwaukee and what happened during the war.

"This is my second visit to Milwaukee. The first time, thirty-five years ago, I had an exhibition fight, ten rounder with a local guy. Can't even remember what the charity was, but for a relatively unknown local, he put up one helluva fight. Don't remember *his* name either. Thirty-five years will do that to your memory, kid. Then again, maybe it was some of the punching that helped.

"Jim," and he held out his hand. It was huge.

"Don." My hand was lost in his big grip. "I had an uncle who fought in Milwaukee: Joe Auchter. Never was knocked down but didn't make the big time. What name did you fight under?"

"My own, Jim Braddock." He said it so matter-of-factly that it took a few moments to sink in. "Oh, my God!" I cried out. Someone silenced me with a hush from the booth above. "You won the world heavyweight championship from Max Baer. I didn't see the fight. I was too young, but I heard about it in the family over a hundred times—from my uncle and my dad. My uncle was the one who did the exhibition fight with you. That was his Milwaukee claim to fame as a boxer. That was years before your Max Baer fight. My uncle fought under the name of Joe Davis. Had a big string of knock outs; never was knocked off his feet but lost two fights by a split decision and he decided to get out. The pride of the family was Uncle Joe's exhibition fight with the famous world champion, Jim Braddock."

I was stunned. The guy sitting next to me wasn't just any old fighter. He was *the* Jim Braddock, former world champ.

Donald Haack

"Yeah, you're right. It was Joe Davis I fought that night. I often wondered what happened to him. He had lots of possibilities, and he had a punch like a mule. Hey, it's a small world isn't it?" And he punched me on the shoulder and then put his arm around me. I was still in a shocked state. It was a small world.

"Guess they're calling me in first. Tell me what you're doing here and what you're talking about." We had a couple of minutes. I filled him in as best I could. "Wish I could hear more about your diamond stuff. Sounds like something I'd liked to have done," he said.

He stood up, shook my hand. "Do well in there and good luck in your diamond hunting. Say, is Joe Davis still alive?"

"He sure is. We're having a family get-together in a couple of weeks."

"Give him my regards and tell him I'm glad it was only an exhibition fight. He was good."

The interview went well and the next day more inquiries were pouring in for Jan to sort through and answer.

After the interview with Jim Braddock, I couldn't help but think about Uncle Joe. He was, as the community termed him, a prizefighter and had a long string of wins. And he was never knocked down. Dad, who thought Uncle Joe was the greatest, used to tell stories about him and enjoyed being with Uncle Joe whenever he had scheduled fights. Some of the tales were about those not scheduled—some unlucky bully who picked a fight with the wrong guy. I also discovered that *my Uncle Joe* was not my uncle, but dad's. Joe was my great uncle, a fact I had a hard time getting used to.

He and I always got along well. He never talked to me about his fights except the Braddock one. But then he didn't have to, Dad told them all. Uncle Joe called me *Happy*, the nickname my grandfather gave me. He and Grandpa were the only ones to call me by that name, and it made me feel we had a close bond. I also wondered if I had some of his traits. I didn't envy the boxing; in fact I really didn't like it, but it reminded me of two fight incidents I had that would have made Uncle Joe proud of me but had the opposite effect with me. I didn't have any grand delusions about them— I just didn't like fighting.

The first one occurred at our downtown Milwaukee YMCA. I was fourteen years old, two years younger than the minimum age to work out in the "Y" at night, but I was tall and there was no reason to question my age. My brother introduced me to the Y before he left for the Marine Corps, so I got to know the personnel and they got to know me. When I started to come on my own, there wasn't any question of eligibility. Every Tuesday and Thursday I took the bus downtown where I went through the routine of running around the track, swimming, and playing with weights.

Then one day I was invited to punch on the bags with the big guys. They put training mitts on me so I could punch the big bag without hurting my hands. They showed me how to coordinate punches on the small hanging bag. It looked easy but turned out to be nearly impossible until they told me to concentrate on the bag a full swing ahead—in anticipation—before it came back. It was too fast to hit when it was coming at you. With practice, I finally mastered battering and controlling it with both left and right hand. The sound was like music to my ears—the steady staccato of beats that reminded me of the drummers in the Drum & Bugle Corps. After I got to know the guys better, I was invited to put on the big gloves and spar with some of them who were aspiring to go on to Golden Gloves, and if they excelled there, on to the professional fights. It was the great hope of many young men, but few would make it.

They were good to me, though. They took the time to show me a lot of the tricks of the trade: staying on your toes, never going flat footed, the importance of weight shifting and how to use your weight to advantage when you hit. They were particularly careful, for which I was thankful, never to outclass me with some opponent who was not in the same weight and experience category. The unwritten rule was: if two fighters were mismatched, the more experienced one never, never took advantage of it. It was a learning situation and not one where someone got hurt.

Donald Haack

I had to cut back somewhat on my running, weight lifting, and swimming when I spent more time in the boxing ring. I liked the people there, and it was the only place I knew someone on a one-to-one basis.

Everything was fine until I missed a Thursday and decided to come in on a Saturday. I did my usual exercises and then went up to the ring where it was more crowded than during the week. There were few people I recognized, so I headed downstairs for a swim and shower, when someone called my name. He was one of the older crowd I didn't know well, but he invited me to join in for a couple of rounds. I was a bit reluctant, but he coaxed me to stay.

When the present round ended, they asked me to spar with the next fellow, a tall, lean black guy whom I'd never met. I should have said no, but I didn't. *Mistake.* The gong sounded and within seconds I knew this wasn't good. I was outclassed by a mile. The real problem was this black guy wasn't playing by the rules we had always gone by.

There was no question about it. He was in a different league. The first round was long and not fun. I was getting tattoo punched around the face with his vicious jab and I didn't have the experience to defend against it. My face and nose hurt. It was painfully obvious we were grossly mismatched. At the end of the round, I took out my mouthpiece and mentioned this to the guy standing alongside. He nodded, seemed to acquiesce and went over to the other corner where the three of them, including my opponent, chatted for a few moments. He gave me the thumbs up sign as the bell sounded for the second round.

Instead of getting better, it got worse. My opponent came rushing out and the right jab was even more aggressive. When I avoided it, I got hit with a jarring left hand to the head, again followed by right jabs. My defensive punches were ineffective against his well-timed, coordinated jabs and punches. This guy was enjoying this with some kind of masochistic pleasure, and I was on the wrong end of his agenda. It didn't seem like I was going to get any help from anyone around the ring. For some reason they either were oblivious or were enjoying the slaughter.

Donald Haack

Halfway through the round, there was something wet on my face and cheek. I wiped it with my glove—it was blood from my nose, which at this point was feeling numb. This fight should have been stopped. My opponent or the referee should have followed protocol and stopped it. No one did. Anger got my adrenalin shooting. I knew then that no one was stopping this fight, and I didn't want to be a punching bag with a battered face. I wasn't in the same class, but I might be able to play dumb and outsmart him. I already knew the sequences of his jabs and punches and was able to anticipate them only to a small degree but not enough to stop the onslaught. He had a pattern of three or four jabs until I defended against them. Then came the left hand punch, which I clumsily anticipated enough so it didn't land solidly. If it did, I would have gone down. He wasn't pulling his punches in the least. I couldn't respond quickly enough. It would only be a matter of time before he connected his haymaker with my head.

I went flat-footed, a sure sign that a fighter was hurt or slowed down enough not to be able to defend himself. It's the signal to go in for the kill. No fighter who is a threat would get off his dancing toes and go flat-footed. I even stumbled a bit to make it look worse and flailed my hand out ineffectually to complete the picture of defeat. Still no one stopped the fight. I was mad as hell.

He came in with his jabs again. I did the usual to stop it, and I could see him lining up his left into my face. This was the one that would put me away. I understood his pattern and knew it was coming. I ducked as it as it zipped over my head. It was the haymaker that would have taken me out. Already back on my toes, I twisted to the right while pushing off with my left foot for added leverage and came around with all my weight to plant a left punch into his rib cage as he hung over me.

I could feel it go in and take effect. All I could think of was hitting the big bag with all my might and weight. This was the hardest body punch I ever threw. I heard the wind go out as he staggered back, totally surprised. He was leaning slightly back from the blow that snapped his head to the right, almost as if to see what hit him. His mouthpiece was half out.

Donald Haack

This is where my fast bag practice paid off. The rule: anticipate, don't react. I knew where his head should be and in the millisecond after the blow to his ribs, I shifted weight from my left foot to the right and lunged at him, pushing off with both legs. The legs were attached to my hips, my torso, my shoulder, arm and glove. It wasn't the ten pound hand-in-glove. It was my full 145 pounds that crashed into his skull. My strategy of anticipation paid off and caught him full in the face between his ear and chin.

If I had reacted a half-second later, he would have been back in balance, tucked his head in and had his gloves ready for the next punch or he would have struck out blindly. But he didn't have that half-second. The momentum flung him across the ring and onto his back. My forward motion was checked slightly by the impact to his head, but I still went forward half-skipping into the ropes. I pushed on the middle rope and climbed down.

"Take these damned things off, you son-of-a-bitch!" I screamed at the guy who did the thumbs up before the round. I spit out my mouthpiece. With my bloody face and yell, I must have come across like I was going to kill him, which I would have if he hadn't hurriedly pulled the gloves off. I glanced back at the ring. Ace was unsuccessfully trying to sit up, his hands were not coordinated and his head rolled from side to side. He was still out of it. They tried getting him up.

I raced down to my locker room and took a thirty-second shower to clean off the sweat and blood. Dressed but half-wet, I was on the street in minutes. Between pressing the handkerchief and applying pressure to the bridge of my nose, I stopped the bleeding just in time to board the bus back home. That was the last time I went to the Y.

A couple of days later, Johnny, the big Italian in our class who was at least two years older than any of us, came up to me. He had a big grin on his face. We never had much to say to each other, and I wondered what instigated this sudden show of camaraderie.

"Man, I just found out what happened down at the Y. Everyone's talking about it. Wish I woulda been there to see it. You knocked out Joe Jackson, last year's Golden Gloves district champ. The guys down at the gym want you to come back." His grin was even bigger now and he put his arm on my shoulder.

"They can all go to hell, as far as I'm concerned. Their sportsmanship stinks, particularly that Joe what's-his-name. He's a real jerk, and I'm sorry he got to be a Golden Gloves man. They shouldn't have let him in." I walked away.

"You dislocated his jaw, ya know," he shouted after me.

Good, I thought. *The bastard deserved it.* Seconds later I was sorry I felt that angry and understood why I didn't like fighting. In spite of liking Uncle Joe, I hoped I didn't have his killer-fighter instincts. I didn't think I did.

The only other incident I could recall similar to the *Y* occurred in the Marine Corp Boot Camp, Parris Island, an incident that could have put me in a lot of trouble.

We were in our last two weeks of our three-month training and the atmosphere was definitely more laid back. It was a Sunday afternoon and some of the men were playing baseball in the area between the Quonset huts, the same area I was walking along to get to my cubicle. The batter came up so I stepped aside to avoid their third base line, which was right in front of our drill instructor's Quonset hut. The batter hit a foul ball, which went over my head and into the quarters behind me. Brian, the big burly assistant Drill Instructor went to fetch it.

Next thing I felt was a heavy impact to the back of my head. Everything went black. Gradually I saw daylight. My head hurt, my face stung. There were tears in my eyes. I was totally confused. I had no idea where I was or what I was doing. I struggled to my knees. My face had been in the dirt. I touched my nose. Tears and mud were streaked on my fingers. My nose ached. I raised my head to see where I was and gradually understood. But I didn't know why I was knocked down. What the hell hit me? There were no cars in this compound. What happened?

A couple of the players came over to help me to my feet. They were holding me up but their attention focused on something behind me. I turned around. There was our burly DI with a bat in his hand, laughing. Laughing at the *home run* he just hit off my head. He thought it was a big, funny joke. My eyes were still blurred from the tears as I tried to comprehend his stupid antics. My adrenalin kicked in before common sense did.

I lunged at him, concentrating on his big fat mouth with that damned grin. I would wipe it off his face. He put up the bat reflexively. I pulled it down with my left hand, lifted off both feet, and landed a right hand to the bridge of his nose between his eyes, as he instinctively ducked, enough for me to miss my target—his grinning mouth.

The blow knocked him back into the hut, over a cot, and alongside the file cabinets, which he almost pushed over with his 250 pound frame. He came to a stop against the tilted file cabinet, his feet up on the cot. He wasn't moving. His eyes were open, the smirk wiped off his face, and his jaw slack. It was a tense moment. Several of the players stepped forward and in front of me—either to make sure I didn't have a second go at Brian or to keep him off me if he came around quickly. It was common knowledge that a private in boot camp striking a drill instructor was the same as striking an officer in the regular corps. It could be brig time, dishonorable discharge, or both.

Reggie, our sergeant in charge and the head DI, quickly took over the situation. He ordered the rest of the men out, told me to stay, closed the doors to the Quonset hut and waited until Brian was back on his feet and thinking clearly.

"Do either of you two assholes have to go to sick bay for treatment?" He waited for us to answer. We both shook our heads no. "Good, then we don't have to go through an official report on that. Do either of you want to press a formal charge against the other?" Brian and I glanced at each other, and then quickly back to Sgt. Reggie. Again we both shook our heads. "Okay, then this goes no further than right here, and we drop it. You both acted stupidly, but it isn't the end of the world. I want you to shake hands and forget about what happened here. You don't have to love each other, but its one helluva lot better than the alternative: inquiry, court martial, brig time—any or all of that. We've only got one-and-a-half more weeks to get through. Let's not screw up." Brian and I shook hands. Surprisingly we didn't have any more problems.

After my head cleared completely, I realized what could have been. I owed Sergeant Reggie big time. I wondered if the event that took place a week before at the rifle range had anything to do with it. The old saying, "What goes around comes around." The last of *that* incident ended up with Reggie saying, "Thanks, I owe you one." It didn't cross my mind until now that that episode was bigger than it seemed at the time. It made me think back about what happened

It was the week on the rifle range. We lived in tents at the far end of Parris Island, where we had to qualify with our M1 Garand rifles. The first couple of days were easy. Those of us who handled weapons before qualified quickly and were awarded our medals. During that same time, the neophytes were instructed on the hows and whys of the weapons and how to shoot. Everyone had to qualify or they didn't make it out of Boot Camp. The last couple of days the pressure was on, particularly for the last eight who hadn't yet qualified. We had to be 100 % qualified as a platoon, which meant those eight had to pass. Some of the instructors and the better shots in our platoon were helpful but others made life miserable for those left. Johnson was one of the unqualified and had two left hands when it came to rifles. We wondered if we could ever get him to pass. One by one we got the others through, but Johnson didn't make any progress. The pressure on him didn't help, it hindered. Several of us who qualified early had spare time and tried to help him through.

The next to last night we turned in at the usual early ten p.m. and were soon sound asleep. Being on the range was much better than in those steaming metal Quonset huts. This was like being in the country compared to the dusty, hot drill area.

At one o'clock in the morning, all hell broke loose. Someone came to our tent and shouted, "Get up, get dressed, and fall in— immediately, and I mean immediately. You got one minute. Let's go." I could hear the shouting up and down the rows of the tents and wondered what could be the reason for this untimely wake-up call. Obviously something unusual. We would find out soon enough. Surprisingly, everyone was out and lined up in minutes, some still tying their shoes or buttoning up their pants. Sgt. Reggie was standing in front waiting for us to line up.

"Get in line, you guys. Button up later. I want you all to move out past the last tent and wait further orders there. On the double, go." Sgt. Brian was standing behind him. It was then I noticed they were both wearing .45 cal. semi-automatic pistols on their web belts. First time I had seen that. Reggie motioned me over. I didn't say anything but there was an obvious question he was going to ask. He motioned me to wait behind him while the troops ran out on the double.

"I was going to ask Greene, the other platoon leader, but I thought he might blow it. Let me tell you what's happened. One of the guys in your platoon went berserk. Took his *pig sticker* (bayonet) and tore up the tent he was in and then ran in circles yelling and threatening to tear up the whole place. He's quiet now holed up in one of the tents, but we don't know which one so we're going to have to search every one until we find him."

"And then what?" I asked.

"Well, we don't know what's going to happen. He's armed and probably dangerous. That's why we're carrying our .45's. And we've got *our* pig stickers, too."

"Do you know who it is?"

"Yeah, we're pretty sure it's Johnson. His tent mate came running up to us the same time we heard the commotion and told us about it. Seems Johnson woke up in the middle of the night with nightmares that became real. That's when it all started."

We waited until the last man was out.

"Can I make a suggestion?" I asked tentatively. I didn't know how my advice would be taken, particularly when there were high emotions involved. It would take a good argument.

"What? What's on your mind?" he said as we saw the last man go out of sight.

"Johnson is out there with a knife, hiding in a tent and probably scared half to death because he's alone and he's fighting whatever demons he had in his nightmare. You're going in there to disarm him and bring him out, right? Someone probably will get hurt. If he's really paranoid he could come out swinging his bayonet. If it's threatening, it could mean someone might shoot him in self defense. Johnson's been under a lot of pressure to qualify. Half the guys have been on his case about it. He hasn't gotten any better and there's only two days left. It got to him and he cracked. He did exactly what he's supposed to do. He didn't get it wrong." I let that sink in.

"I don't get your point. Whaddya mean, he's doing what he's supposed to?"

"It's the system. Boot Camp is programmed to make everyone work as a unit, not thirty separate guys running around. Boot Camp is supposed to break those who don't make the grade. The philosophy is it's better to break in Parris Island than on the battlefield where a guy may lose his own life and endanger the lives of those around him. We weeded out five in our platoon the first three weeks. Pissing in bed was one of the symptoms for a medical discharge. When some of the guys heard that, they faked it. That's why the doctors used those oversized dull needles to inject salt solution into a guy's ass for bed-wetting. The fakers were eliminated immediately. No one would voluntarily undergo three of those shots. Those that got all the shots didn't have a choice—they were chronic bed-wetters, caused by the stress of Boot Camp. And they got their medical discharge."

"Johnson didn't have the option of pissing in bed. He just broke under the strain of not qualifying on the range and felt that he let the platoon down. He doesn't like guns, he's under pressure, and he did exactly what the system wanted him to do—he broke down here, rather than on the battle field. He's not a bad guy. I don't think he would hurt anyone if he could be talked to. He's from a farm in eastern Carolina where he worked tobacco with his family. Before this blows out of proportion and someone gets hurt, would you give me a chance to find him and have a talk? We got pretty close this last week, and I believe he'll listen to me. What do you say? Give me an hour after I find him?"

Donald Haack

Sarge waited for what I thought was a long time and then twisted his jaw back and forth as if trying to make a tough decision before he said, "Okay, I'll go along with it. I shouldn't, but you've got a point—the recruits *are* supposed to break here not under real fire." He unstrapped his web belt that held the .45 Colt to hand it to me. "For God's sake don't use it unless you really have to. We'd have to answer a lot of questions, and I'm going out on a limb."

"No, no, I don't want the pistol."

"Okay, then take the pig sticker. At least you'll have something to defend yourself with."

"No on that, too. Nothing. But could I get Raul to come along with me? Raul was pretty sympathetic to his case and Johnson would feel comfortable with both of us there."

"I think you're crazy to go in there unarmed, but if you think you can pull it off, it sure would be the best solution. Okay, I'll get Raul back here."

Raul was my old Alabama chain gang buddy who I was teaching to read and write so he could stay in the Corps. In his former life, in one of his drunken moments, he stole a train and ran it through three states before it ran out of fuel. He served two-thirds of his sentence in the Alabama chain gang and got off for good behavior if he enlisted in the Marines and stayed on active duty. Reading and writing is a prerequisite for a Marine. He had a great incentive to become literate, the alternative going back to the chain gang.

I quickly explained to Raul what the situation was. "We'll start where his tent was cut up and then do an ever-widening circle and hopefully find the tent he's hiding in. I don't think he'd run far after his first cut-and-slash." Raul agreed. There wasn't a moment of hesitation when I asked him in front of Sergeant Reggie if he would voluntarily go with me.

We found the tent Johnson decimated. He must have been in some really wild hallucinations when he tore it apart with his bayonet. I quietly hoped I hadn't made a mistake in guessing what emotional state he was in, and if in fact he was even able to understand what we were trying to do. We immediately searched the six tents around the damaged one. Raul opened the flaps as I called out Johnson's name and flashed the light around the inside. I didn't want to surprise him and have him react unpredictably—and possibly attack before he knew who we were. We extended our search to the next ring of tents. That's where we found him—crouched down in the far corner of the third tent, holding a bayonet in front of him in a defensive, defiant position.

"Johnson, it's Raul and me," and I lit our faces with the flashlight so he knew for sure it was us. "We're worried about you. Can we come in?" He didn't answer but nodded yes. We stayed on the far side and bent down on our haunches so as to not to be intimidating by standing above him. I explained that the DIs out there were confused and had guns and that we talked them out of that. "We aren't carrying any weapons and we made sure those guys backed off. Didn't want something stupid to happen and possibly get you hurt." I continued my monologue and Raul interjected a bit of humor about his wild life on the Alabama chain gang, so Johnson could empathize with him—hopefully to understand he wasn't the only one with problems. Raul was good at soothing the tension. Together, we made some inroads.

"Give us your bayonet and then walk back between us to where the MPs will take over. You'll spend the night in the brig tonight and probably get transferred to the hospital tomorrow or the next day. Raul and I'll visit to make sure everything is okay. With any luck, you'll receive a medical discharge and be back on the farm with your father within a month. Hey, and you're not letting down your platoon. Raul and I'll explain. They'll wish you all the best."

Johnson was crying.

"Johnson, toss the knife to me and let's go back." He didn't raise his head, he was sobbing convulsively. He turned the bayonet around and with the handle towards me, threw it across the floor. Raul picked it up and I moved over and put my arm around Johnson. We helped him to his feet and the three of us walked out together.

The MP was there waiting. He cuffed Johnson and maneuvered him onto the jeep.

Donald Haack

"Raul and I'll be over to see you in the morning. Hang in, everything will turn out okay." In minutes they were gone.

Sgt. Reggie came over. "You were right about Boot Camp making guys work as a unit. He was part of your platoon and you and Raul put your neck on the line for him. Maybe the system does what it's supposed to do." He stuck his hand out. "This could've been messy and it turned out fine. Thanks, I owe you a big one."

Mother & Dad at Chicago travel show.

10. THE ERRANT PILOT

Back in Milwaukee I received a telegram from Frank:
Really concerned. Patlow flying Cessna
every day. Bertie says over fifty hours
this week but not one revenue dollar
received. Plane needs maintenance but
he's avoiding me. Advise. Frank

"It never lets up, does it?" I remarked to Jan.

"Well, you can't go back right now. I've got this week lined up, mostly in Chicago. Jack Eigen, WMAQ, wants you on the show tomorrow morning. Sig Sakowicz has booked you for a second interview. That's on Dearborn, not far from WMAQ, so you shouldn't have any problems with that schedule. Then Wednesday, Ed Sainsbury, UPI news, wants you in his office first thing, 8 a.m., on North Michigan. He'll do a news release on the Safari. Two hours later Jerry Harper, WBBM-TV, will tape an interview for their *Seven O'clock Report*. That's on McClairy Street, not too far away. Late afternoon, Bob Elson wants you in the Pump Room, Ambassador East Hotel, for their WCFL broadcast. Later in the week, George Leighton is setting up a group meeting for any of our interested clients. Says we can hold it with the members of their *Adventurer's Club,* on West Jackson. They'll host the whole shebang if you give a twenty-five minute talk and answer questions. Boy, you're getting popular. We average ten potential client calls after every appearance."

Jan was right. Calls were coming in, followed by deposits. Bookings filled in straight through the dry season, unofficially ending mid-April. We couldn't take a chance with getting hit by the rains, so we stopped scheduling after March.

On the weekend we received another telegram.

> Patlow flew Cessna every day and abandoned it in interior. Condition unknown. Bar rumors, Patlow says plane will never fly again. Bertie calculates $9,000 in missing revenue from Cessna. Patlow thief, over-riding greed eliminated any maintenance. Plane is lifeline for mining and safari operations. Urgent you return. Frank.

I cabled him right back.

> Find Clavier or other pilot. Need to know where and what shape 180 is in. If broken parts I can bring down, i.e. prop, battery, starter, generator. Don

I waited two days then received another cable from Frank.

> Clavier found plane north of Kamarang, sitting end of runway, slightly tilted, door flapping open, covered in oil but saw no physical damage. He will fly us in and help with any repairs. Frank

My return telegram:

> Frank, leaving here Friday, ETA Saturday. Arrange Clavier charter for Monday. Need following supplies: Your tool box, battery, bar soap and a spool of thread, 10 gals Avgas, spark plugs, tire pump and repair kit. Get inner tubes, tires and oil filter from Bertie. 6

tins of oil, flat and round file, hacksaw, plenty
of rags, 30 ft bailing wire, fine sandpaper,
clean chamois and funnel, 1 gal empty Klim
tin, plenty matches. Pack a lunch and beer.
Don

Frank met me at the airport with Bertie. "Got everything you asked for. Gotta ask you, though. What the hell is the bar soap, thread, and empty milk Klim tin for? Everything else I can figure out."

"Those are for bush repair tricks. String and soap can repair broken oil and gas lines. Sounds crazy, but it works. Klim tin is the big can that powdered milk comes in and we might have to use it as an oven if the inside of the magnetos are damp. Temporary fix is to heat 'em up for a couple of hours to dry out until we can permanently seal or replace them."

He screwed up his face. "You're not putting me on, are you?"

I assured him I wasn't and would show him a few emergency bush repairs he hadn't seen before.

Two days later, Tony Clavier dropped us in where our wounded bird sat guarding the end of the airstrip. It truly was abandoned. I felt sick that anyone, especially a pilot and aircraft mechanic, would be so indifferent about a plane.

Frank and Tony started by cleaning up the oil leaks so we could see how bad they were, while I tackled the flat tire and the door, which had a broken latch. In an hour we pretty well summed up the major problems. One gas line had a leak and it was a stroke of luck that the plane hadn't caught fire. There was a long thin crack in one oil line that caused the mess over the whole plane and part of the windshield. I drained a couple of ounces of gas out of the quick drains.

"Okay, Frank, this is how you fix oil and gas leaks...until you can change the lines." I cut the thread into a 24 inch piece and wrapped one layer around the oil line crack, then took the bar soap and rubbed it on the thread until it was fully covered with a layer of soap. I used a small brush to wet the soap with gasoline, which formed a kind of gel. Another layer of thread tightly wound, more soap, wet again with gas. "That should do it but we'll overkill and do one more layer. In an hour that patch will stiffen and be better than the original line." I did the same for the leaking gas line with the tiny pin point hole dripping gas.

Frank watched in amazement. "Those patches really gonna hold?" Tony, meanwhile, removed the spark plugs that were chock full of black carbon. *I'm surprised they got the plane this far.* We changed the plugs and the oil—the blackest, dirtiest that any of us had seen. The tire had a small leak, which we patched and filled with the tire pump. The other tire was low but only needed air.

"We do have some work on the prop. Don't know where and what he was flying through, but there are a lot of nicks on the leading edge, which have to be rounded out. They could cause a crack." Tony wanted to get back before dark. Some of the obvious problems could be solved tomorrow, but first I had to find out if the basics worked. We rotated the prop—each cylinder had good compression, a good sign. With the new battery I cranked up the engine. It roared to life. I checked the instruments: oil pressure okay, generator charging, temperature staying in normal. With full RPM, both magnetos were right on. We shut down.

"Tony, we have it under control. Thanks for the fly-in and the help. We'll work on it today and finish up in the morning. Even the radio is working, though the back antenna is loose and about to fall off. We should be back at Atkinson in the early afternoon. Thanks again."

We shook hands and he took off. It's amazing how desolate a place seems when there are no civilized sounds and the scenery goes on forever in all directions without man's intrusion. We directed our attention back to the plane.

At dusk, after a couple of beers and Mama Mittlehauser's sandwiches, we hung our hammocks and within minutes were sound asleep. The next morning we finished at eleven, did our pre-flight check, re-did it twice and were on our way. It was a comfortable and warm feeling to get behind the controls of this dependable workhorse, and I cursed Patlow for abusing it so badly. In another couple of hours, with Frank's magic touch and tender loving care, we would have it as good as new. Our next big problem was Patlow. Not only did he steal all our revenue, but we heard he was disregarding his own Seabee's maintenance, and legally he was still attached to our group.

After Frank and I had our Cessna up and running, my next line of business was to find Patlow, get an explanation from him, and hopefully retrieve eight-to-ten-thousand dollars he received from flying our plane. Considering what he had done already and was continuing to do with his own plane now, I had a pretty good picture of our *new* Jim Patlow; he no longer was the same person we brought down with us.

I'd seen this type of behavior many times before and realized there is no way to anticipate the character change of a person exposed to a drastically different life style, particularly when a greed factor is involved. In combination, the probability of character breakdown is high—a condition that Patlow seemingly succumbed to.

I thought of the professor who came down on a one month leave of absence to get away from the strain of teaching and counseling. After a week in the bush, he no longer combed his hair, stopped shaving, and worse, wore the same clothes without daily bathing. He looked and smelled like something that came from beneath a rock and resisted any form of suggestion. He had to be forcibly removed to a hotel, ordered to bathe and dress in a civilized fashion before he was sent back home.

Then there were the two guys from the States, one an engineer, the other who owned his own equipment rental business. They came down to help a friend who was prospecting for diamonds. They were going to show him how to modernize his methods rather than use the crude native procedures. Cut off from civilization's TV, newspapers, movies, and McDonald's quick lunches, they quickly and enthusiastically fell into the embraces of local rum. They were stumbling and blubbering drunk during the day. In the evening, when not drinking, they fell out of their hammocks and were stung and bitten by insects to the point they were hardly recognizable. One fell into a creek and wasn't sober enough to extract himself. He only survived because an alert Indians saw him and dragged him out. I was to fly these two guys from one camp to another with some equipment and rations for the crew they were to visit. My prospective passengers were too drunk to get into the plane. Moreover, when I realized how totally incapacitated they were, I didn't want them in the plane.

I noticed one had urinated all over himself and wasn't aware of it. The other found it funny, so hilarious he couldn't stop laughing and fell down. We couldn't get him back on his feet. I called the nearby Indians and asked them to carry the two back to camp. I cancelled their flight. These were two educated, church-going, upstanding family men who only had an occasional drink back in the States. They were obnoxious, roaring drunks all the while they were down here.

Living in the bush is a challenge for many adventurers coming from civilization. It quickly became evident why the English, who had hundreds of years of colonial experience, stressed: *Do not go native.* We followed our British friend's persistent advice: have tea in the afternoon when possible, shave and shower every day, change clothes into something proper before dinner and limit the before-dinner drinks.

So Patlow wasn't the first to succumb to the *Colonial maladie.* Our first problem was to find him. He wasn't filing flight plans on his landings and switched between Atkinson Field and Ogle on the coast. Harry Wendt, my engineer friend at Ogle, informed me he had a chance to look over the plane when it last flew in—it was a mess. Apparently, Patlow landed or took off through some thick brush. The prop had all sorts of nicks, some quite deep, from the tip to the hub. Harry said the plane should be grounded. It was a real hazard.

Jim wasn't at his regular hotel. He had moved back and forth in the boarding houses so we couldn't find him. I had enough. When I heard from the tower that he landed at dusk, Bertie and I drove out to Atkinson field and posted a legal notice on his plane that I had a local lawyer write—a termination of all business transactions with him. I enclosed an overdue bill for $9,350 for 105 hours of tach time on the Cessna at $85 per hour to be paid immediately or a lien would be filed on his plane.

I sent a copy of the legal notice to the Director of Civil Aviation, in Georgetown, and two copies to Florida to Jim's parents and his lawyers. Telegrams were sent to Florida advising them of the problem and terminating the relationship, stating that a letter was to follow.

I didn't have time to waste on the Jimmy Patlow problem. Our new safari guests arrived, and we were all kept busy. But we still heard via the grapevine that Patlow was flying his plane without maintenance and there were some serious issues, particularly the prop, which was not cleaned up from the damage it sustained. Everyone agreed that he was an accident waiting to happen. The following week when we received a call from the DCA that Patlow's plane was missing, it wasn't a big surprise.

After the call I received a hand-delivered note from our DCA, Alex Philips, asking me to call him immediately. He had received my legal notice terminating business with Patlow and wanted to know everything about his flying and what caused the termination. I called and was put through immediately, something that seldom happened. It had to be serious. I gave Alex as many details as I knew.

He explained, "Some miners in Kaiteur Falls were to fly two trips with Patlow into a new airstrip about ten minutes south. The miners left behind said that when the first flight was a few minutes out it had engine problems and turned back. Moments later the engine picked up and ran smoothly so it continued on its original heading. It was out of sight when they heard the engine start and stop several times before it stopped completely. They figured the plane was a long way from any airstrip. That news was two days old and there's been no word since."

"Hmmm, that doesn't sound good. Has anyone had a chance to look around up there, yet?"

"No, you're the first pilot I've talked to. No one else flies around that area. Do you know the airstrip they're talking about? I don't have anything listed on my charts and no one has applied for a strip there."

"There's a small open area on a hill above a waterfall ten minutes south of Kaiteur. I saw some work going on a couple of weeks ago and assumed it had something to do with the missionaries at Paramakatoi, a few minutes flying east of there. You might try to contact them to see if they know anything about it. If Indians were working on it, they would've come from there. I'm making a flight to Echilebar early tomorrow morning. I'll take a look and see if I can see anything. At the same time, I'll take a compass heading out of Kaiteur to the strip, in case we have a search and rescue—which seems pretty likely, don't you think?"

"Yeah, I do. What makes this more difficult is that Patlow was illegally flying the interior by not filing flight plans between airstrips. If he filed out of Kaiteur and didn't close at the other airstrip, we would have a red alert immediately. He's broken too many rules lately. If he's alive, he's got a lot of explaining to do if he wants to fly in this country again."

"Let's hope he's alive. He had three other passengers on board. You can expect a call from me about ten tomorrow. I'll route it through BG Airways."

The next morning I picked up three of my safari clients. I explained the situation of the lost plane and that we would do a search pattern on our flight in. Each client was assigned to scan the bush below in quadrants, from below the plane to one-quarter mile out, then the area alongside and scan back and forth again in the same manner. That way we could cover a fairly substantial area. I briefed them to look for any smoke, break in trees, or anything shiny that wasn't water. I checked the time and wrote down the compass heading, which would only be a rough estimate on this first leg. I didn't circle—I needed to establish a correct heading and time to the airstrip, if I could locate it.

Nothing showed up on the first pass, even with my enthusiastic passengers getting involved in this real search and rescue on their first day in the jungle. They were shaking with excitement. I needed help and they responded enthusiastically.

Eleven minutes out I spotted the opening where something akin to an airstrip was being constructed. It wasn't much, but then neither were any of the temporary strips built in the interior. None of them would be approved by our FAA in the States. If those standards were applied here, there would be no mining or exploration done in the country.

I flew over the area. No sign of life. I took a reciprocal heading, added 5 degrees east of my flight out and used that heading back to Kaiteur, which I could see a long ways off because of the water vapor clouds created above the falls. I corrected three degrees, accurate enough for any search planes to use for the path that Patlow must have used. This time I circled several times in the area where his engine might have stopped. Nothing showed up.

I called BG Airways. When I couldn't raise the DCA, I left the information with the operator to pass on. It was 10:15 a.m. If the DCA wanted to do a search, he now had the heading and the time calculated at 130 knots to reach the airstrip of Patlow's destination. There was no plane on or near the strip. It was becoming obvious he had gone down somewhere in the bush below. I advised the DCA I would be available for further searching in the afternoon and the following day. After the fly-over, I was convinced there was a downed plane. I damned Patlow for not following rules and not calling in his flights. He might be flying somewhere else in the interior, and we're about to waste more gas, time, and risk lives for this irresponsible joker.

I checked back on the radio every hour. It was two o'clock when Alex Philips called. He, too, was convinced we had an air crash somewhere up here, and it was critical to find if anyone survived and required help. He was trying to contact Cominco, the subsidiary of the Canadian Pacific Company doing exploratory work in the interior— they had the only helicopter. He asked me to continue the search. We were the only plane in the area.

I explained this to Frank and our safari clients. They were caught up in the excitement of a real life bush crisis, and they all agreed I should go. Frank and one of our clients, Larry, who had done some search and rescue in Canada, volunteered to come along.

We flew over the area but by then the sun was low in the horizon, making it difficult to spot anything in the tall trees below. We decided that it would be easier to spot a hole in the canopy with the sun overhead between 11 a.m. and 1 o'clock. We returned to camp that afternoon without any results. The next morning after talking to the DCA, we explained what we were going to do. We had enough fuel at Kato to fly a couple of hours with the mid-day sun overhead. Even though the area we pin-pointed was relatively small, it was like looking for a needle in a haystack.

If anyone were alive, I was hoping we would see some signs of smoke, either from a crash or more hopefully from a signal from the survivors down below. Flying low and slow over the tree tops yesterday should have revealed any telltale smoke. There wasn't any.

Today, we planned to search directly below our plane in the hope we could spot damaged trees or more likely spot the sun's reflection off the plane's fuselage. I checked my watch. One hour and fifteen minutes over the grid with no results. I set up another imaginary grid this time closer to Kaiteur. We were one-third into that grid.

"Don," Larry shouted. "I think I spotted something shiny down there. Can you do a 180?" I came around to approximately where Larry called out. We circled. Nothing.

"Keep circling. I'm sure I spotted something shiny, but it was just for a second. Man, the bush is thick down there. Look! Look! There it is again—shiny and what appears to be the white of a broken tree."

I pulled a tighter turn and it took two circles before we all spotted it, but only for the slightest fraction of a second. I kept circling where they spotted the hole, while I triangulated points of references. We were on a south heading of 185 degrees from Kaiteur. To the east, the tallest mountain lined up at 95 degrees east. Approximately five miles west at 260 degrees, was an east-west bend in a river, probably the upper Potaro. That gave us a fairly accurate triangulation for the helicopter, if it came. Otherwise it would be days or weeks before any rescue team could reach the wreckage on foot. I called in the information to Alex and returned to our base camp at Echilebar. It wasn't speculation anymore. We had a confirmed air crash and a somber mood in camp. The only speculation—were there any survivors?

Cominco regularly called in on all their interior flights, which made it easy for air control to contact the helicopter pilot. Within an hour the Cominco helicopter was winging its way to Kato, where I had my closest cache of aviation gas. I met them there. They refueled and we discussed the coordinates. We agreed I should fly to the crash site and circle, eliminating the need for the pilot to calculate the coordinates which would use up precious time and fuel.

We felt the best plan was for the helicopter to use their cable winch to lower one of their men with a radio, first-aid kit, and chainsaw into the small hole in the tree canopy and then release the cable, so it wouldn't snag on anything that would endanger the helicopter. We flew in a circle far enough away so we wouldn't interfere, but close enough to keep radio contact in case we saw something go amiss.

Getting a man down the hole was like threading a needle. He kept hanging up on the tree limbs alongside. The hole was impossibly small. They hovered and maneuvered until he was on the ground where he unstrapped his equipment and signaled to pull up his life line. He was on his own. He was to check for survivors if any, treat them and then report the condition of the crashed plane.

The helicopter hovering directly above had a difficult time seeing what was happening. The pilot could see a reflection of metal but couldn't make out the wreckage itself. The helicopter didn't have large reserves of fuel so I radioed to them to go to the airstrip a few minutes away and wait for us while I kept in contact with their man on the ground.

His first report was grim. The plane was a compressed crumpled mass of aluminum. There appeared to be three persons inside, crushed badly, and no sign of life. A fourth person was thrown out through the roof of the plane and might have survived. There were open tins of food and a canvas shelter that he must have salvaged from the back of the plane. The Cominco man shouted in all directions, but no one answered. He radioed back to us that the smell of gas was strong, and he was afraid of doing anything that would cause an explosion. He was surprised the plane hadn't caught fire.

Patlow's Seabee apparently came straight down and sheared off the tops of a couple of trees. Other branches spread out to block the view from above. We were extremely lucky to have spotted the fuselage. Cominco's man on the site tried to enlarge the hole with the chain saw. The trees were so thick that when he cut two or three they remained standing, held up by the heavily-clustered surrounding trees. By cutting them into pieces, they came down lower and finally fell to the ground. The helicopter came in and pulled their man out. I radioed Alex Phillips the details of the accident scene. A ground crew would have to go in from the small airstrip site. They could use auto jacks to open the fuselage. Cutting torches would set off an explosion. The bodies could then be removed, carried to the airstrip, and transported to town. There was still the question of the fourth passenger, who appeared to have survived but disappeared.

The next day a ground crew was flown in to the nearby strip, and three days later they reached the crash site. The same day a porkknocker (the term used to describe the miners in the interior who had a diet of salted beef in barrels, and who had to pound on the barrel to remove its top, a sound that could be heard throughout the mining area) arrived in Madia, the old mining site below Kaiteur Falls. He apparently was the survivor from the air crash and was in critical shape, having a badly infected compound fracture below the elbow. He had numerous cuts on his head and chest and was suffering from dehydration and lack of food. To be in that terrible condition and be able to walk out that distance in the bush was a miracle in itself. He was made of tough stuff but incoherent when the Indians found him. After two days in the hospital he gave an account of what happened leading up to the time of the crash.

He couldn't remember being thrown out of the plane, waiting around the wreckage, or his seventy-five mile trek to Madia and how he even found it.

He remembered when the plane first had engine problems. "We told pilot we want to go back to Kaiteur— right now— and he turn back. Engine sound bad and we think we not make it. Then engine sound good and we were happy—until he turn plane back to small strip. We shout, 'Back to Kaiteur!' Pilot shake his head, say it okay… just dirty fuel that went through—no problem. Then engine run rough again and then it stop. He say he going to land at new strip. But it too far away. We get slower and slower, the trees closer and closer. Then plane turn over and drop straight down."

We all understood what that was: the plane stalled.

The bodies arrived in Georgetown a day before Patlow's parents arrived. They wanted to see me to find out what happened. The U.S. Consulate's information was sketchy. It all happened so fast. The Patlows and their lawyer had received my letter a few days after my telegram and were confused. Less than a week later, they were informed by the State Department about Jimmy's crash and death.

The night they arrived they were stressed out and in a shocked frame of mind when I met them. Mr. Patlow was talking to the consulate and Mrs. Patlow pulled me aside, asking for all the details, starting with the reason I separated from Jimmy.

I gave my condolences and explanations as diplomatically as possible. I informed her about the telegrams from Frank Hawk when I was in the States. She nodded and reminded me she met Frank when we stopped in Florida to pick up Jimmy. Without sounding judgmental, I gave her the facts, toned down to make them sound less harsh. I explained that Jimmy had abandoned my plane in the bush but I omitted the fact that he owed us a great deal of money from charters. It was hard to give the facts on his last two weeks of flying without making them sound offensive.

It didn't matter what I said—not *her* Jimmy. He wouldn't do anything like that. He was a good, hard-working boy. She was in denial and I wasn't going to convince her otherwise. I told her to talk to the consulate. They had all the facts. I was hoping to extend my condolences to Mr. Patlow and then leave, but he wanted *all* the facts. I tried to explain that I gave most of the details to Mrs. Patlow, but she really didn't want to hear them or believe them.

"I understand what you're saying, but do me the favor of telling it all to me again. Only this time don't leave anything out that you probably did to spare Mrs. Patlow's feelings. I really want to know what happened—all of it."

So I told him the whole story from the beginning of Frank's telegram.

"How much money did he make for the flying in the last month?"

"We don't know. He never reported it to Frank or Bertie and avoided me when I returned."

"How much do you think it was?"

"At the going rate of what we charged, between eight and ten-thousand dollars. Frank heard from his friends in Georgetown that he needed twelve-thousand to buy his own Cessna 180, and he was about to go back to Florida within a week or two. Apparently he had that much cash on him."

"What do you think he did with it?"

"From what we heard, he kept it on him or the briefcase he always carried. He was afraid of being robbed if he left it in town and didn't like the idea of using the bank. He didn't want a record of what he had."

"Where do we go from here? It's obvious he owes you money. Was any cash found in the wreckage?" I explained to him that there was no record of any money found, and I didn't think there would be. Cash has a history of disappearing in bush accidents. I told him that if any cash is found, I would like to have part of it. It belonged in our business and many people were dependent on it. However, under the circumstances, I was not going to file a claim against the estate. The Patlows were going to have serious expenses, including shipping their son back to the States. They were not wealthy people. It turned out that I was right in my assumption—no cash was ever found.

Sometimes being a parent can be very difficult. Being a partner can also be difficult.

11. THE MAD RUSSIAN

At the end of the dry season after our tourists left we operated two portable dredges on the Ireng River, the border between Brazil and Guyana. Our camp, ten miles north of the Echilibar River, was along an airstrip I built years ago for Jimmy Angel. It was beautiful open rolling country with a 2,000 ft. high escarpment forming a backdrop to the east. There were patches of thick jungle in the valleys dotted by swamps and creeks that meandered in an irregular pattern. On a clear day we could see Mt. Roraima bordering the third country, Venezuela.

Mt. Roraima, a formidable two-mile high plateau of sheer rock that sprouted out of dense jungle, was best known as the site of Sir Conan Doyle's *The Lost World.* The natural massive fortress of Roraima and the desolate wilderness provided an appropriate setting for mining diamonds.

After two months we had a first-class airstrip and our camp consisted of two sleeping tents, one storage tent, and two more for cooking and eating. Not complete comforts of home, but not shabby, either. We had one month of dredging before the rains started. After that it would be too dangerous to work the treacherous swollen rivers.

This was our lucky site. Our first strike had been a 7½ carat diamond, followed by a 4 carat and several 3 carats. We were "hitting" diamonds every day in relatively shallow waters and each sluice box of gravel produced a half-carat or more.

Donald Haack

About the same time, activity across the river picked up.
Mules were coming and going and Amerindians were drugging
(local slang for Indians carrying goods on their back) in supplies,
highly unusual for this remote area. A few days later, a tall guy
with a black bushy beard approached our camp, shouting, waving,
hands spread high in the air, and smiling from ear to ear. Even
though he was bearing two pistols and a bandolier belt across his
shoulder—not too unusual for a Brazilian in this territory— he
appeared to be on a friendly mission.

Our firearms within reach, we cautiously waved him into
camp. There was no law and order here, two-hundred miles from
the nearest town. Weapons were the only *law and order.* Everyone
carried a gun. In broken Portuguese interspersed with Russian,
he told us his name: Attanasoff. He came from Russia and lived
in San Paulo several years before drifting north to Manaus on the
Amazon. From there he continued north up the Rio Negro and Rio
Branco to Boa Vista where he heard about diamonds and decided
to prospect for them.

He spoke with an intensity close to madness. He flailed his
arms in big circles for emphasis. His eyes were bulging, wide
and flashing, but he seldom looked directly at us. His beard and
untrimmed mustache were blended together and framed bright
red lips constantly flicking out saliva when he talked. A red
bandanna around his head partially covered hair crawling out in
all directions.

We addressed him as Gospodin, Russian for *mister* and Senor
Attanasoff, but when he left, he became *The Mad Russian.* He
had been dredging downstream for months but had trouble getting
diesel fuel which had to be transported by mule in five-gallon
jerry cans from Boa Vista, a two-hundred mile, seven-day trek.
Supplies in Boa Vista regularly ran out.

He heard that a *gringo* with an airplane stockpiled fuel from
Georgetown and helped Brazilians who were in need. He said
this smiling broadly, swinging his arms in wide arcs and spittle-
spraying anyone within five feet. He pointed to me to emphasize
his wild praises. I should have kept my mouth shut, but my ego

over-ran common sense and I spoke to him in Russian. Mistake! He declared we were blood brothers. He actually did a mini Russian *Satchka* dance, he was so elated to hear his language spoken. I studied Russian at the University of Wisconsin along with an advanced summer session at Middlebury School of Languages in Vermont. Never did get to use it in the service, and I certainly could have done without it in South America.

I explained that my first priority was to supply my camp. I had one extra 55-gallon barrel of diesel I could sell him now and could bring in another barrel for him in the next couple of weeks. That was more than he expected. He was eternally grateful, he said. It was worlds better than bringing it by mule in 5-gallon containers as he had before.

In the next four weeks I brought him two barrels at one-third the price of what he previously paid. His good mood, however, lasted only as long as the diesel fuel. When it ran out he became nasty. His dredge was situated a mile upstream, but his camp was directly across the river from us. Their crew of eight were on the dredge all day, as we were, and we only saw their campfires at night. Both of our dredges worked long hours and double shifts, which is what you do when working good diamond deposits. In this area we had to remove three feet of gravel overburden to expose the diamondiferous gravel. The yield was good: a carat of diamonds every few tons. We worked every daylight hour and into the night to ten or eleven o'clock.

The late hours made our foreman, Frank Eagle, disgruntled. He wanted everything done by the numbers and didn't like to have his precise routine disrupted. I played referee between Frank, who wanted everything shut down at 5:00 p.m. before dark, and the eager divers, smitten by diamond fever, who wanted to work around the clock. Good strikes, few and far between, raise the tempo and everyone in camp goes into a fever pitch. With everyone sharing in the diamonds, long hours were the norm.

The men on our dredge previously earned fifty dollars a month in their former jobs as laborers, bartenders, and taxi drivers. Amerindians earned far less. A dedicated diver in a few good weeks could earn the equivalent of what they would make in ten years. For that bounty they were willing to risk their lives and put in long hard hours. Erratic eating and sleeping habits created frayed tempers. Sometimes it took an act of God to keep a lid on the numerous squabbles to keep the camp from blowing up.

It was in the middle of this scenario that our *friend*, the mad Russian, also started finding diamonds. He operated his dredge through the night, using up precious fuel at an enormous rate.

He believed it was his God-given right to be supplied by me— and for me to drop everything to be at his beck and call to haul diesel and food to his camp. My explanations as to how much time I spent flying out of camp fell on deaf ears. We knew there was a serious problem when he arrived in our camp, armed to the teeth, belligerent, demanding and threatening— with an ultimatum. If he couldn't get fuel, he would shut down our dredge. He stomped out of our camp and into his boat. We didn't know how crazy or serious he was.

We had our answer the following morning. Gunfire awakened us! Shots were coming from across the river. Two tents were hit, but fortunately no one was inside. It sounded like he had a small army and plenty of weapons. Over fifty shots were fired. We quickly moved the diesel barrels out of range and pulled the airplane to the far side of the hill where it couldn't be seen from the river. We carried lanterns, stoves, and critical supplies out of the tents. Downstream and around a bend our dredges were safe from even his wildest shots.

We needed a strategy and quickly. We didn't come here to have bloodshed or wage a war. Even though there were no police around, it didn't mean there wouldn't be an international incident with police or military being brought in if someone were killed. It was clear that Attanasoff wasn't about to go away. He apparently had plenty of firepower, but Amerindians were poor shots. Bullets were expensive and scarce, which made all the difference when accuracy depended on practice. Frank and I were both good shots, and we had plenty of ammunition...but we didn't want a war. We made the decision—we had to be on record with BG authorities to declare Attanasoff as the aggressor— before the situation deteriorated further.

Shooting in self-defense against a known aggressor is more acceptable than an undocumented shooting war, which this would be without outside observers. Meanwhile we needed to restrain Attanasoff and keep him off balance while we filed a report with authorities. To do that we needed his full attention.

I had an idea and discussed it with Frank. He agreed. Frank's eight-power telescope on the .22 caliber Hornet center- fire rifle came out of its bag. We took pride in our regular target practice to sight it in to hit *a gnat's ass*, as Frank termed it.

We set our plan in motion. Climbing to the highest knoll, we stayed behind scrub bushes and scanned Atanasoff's camp with our binoculars. Their storeroom, a crude mud shack with an opened fold-up window, revealed bottles of rum and tinned goods on two shelves. Lying in the sun concealed in brush, we propped the rifle on a solid log and made ourselves comfortable.

At eighty yards it took two rounds to sight in and then the fun began. Frank shot. Using the binoculars, I called the hits to adjust fire. We took out every bottle of rum on the shelf. There was a big ruckus in their camp and everyone ran for cover. Without knowing where we were, they ran in a half-crouch believing they were out of our sight. They saw the store and the bottle wreckage. Shielding their eyes from the sun, they scanned the river, pointing one direction then another. They weren't smiling. There was no sign of Attanasoff.

One of the "enemy" pulled out the stick holding up the store window panel, closed it, and slipped on a lock. That brought a chuckle from Frank. I smiled. Our thoughts were the same. We watched them sneak back into their tent, which they thought was hidden from us, but couldn't have been more exposed.

I nudged Frank. "Maybe they got the lesson, but let's reinforce it."

"Take out the lock?" he asked, taking the words out of my mouth. He glanced at me and frowned. "You're pushing it." I knew

Donald Haack

Frank well enough to know this was just the kind of challenge he thrived on. "Spot me," he whispered. I did.

The first shot went two inches high, the second was one inch out at three o'clock. The third took out the lock. Attanasoff's men ran to the store and saw the shattered lock. They stared towards our camp. We made our point.

We gave the divers one standard .22 caliber rifle and a pistol, with instructions to shoot near but not at anyone who threatened them, but only if they were fired upon. If they felt their life was really threatened they could do whatever they had to in defending themselves. That evening Frank, Domingo, and I pulled the plane onto the airstrip, topped up the fuel, and did a preflight.

An hour before daylight we gave Domingo a storm lamp to hold as a visual guide at the end of the runway. Frank and I climbed into the Cessna 180. I turned the key and the engine roared to life, breaking the stillness of the night with a shock. Wasting no time, we lined the plane up with the lamp, set the gyro compass, and accelerated down the runway. I had to switch glances between Domingo's lamp and the instruments panel, which I would have to use exclusively after I took off into total darkness where vertigo can occur quickly.

Several friends lost their lives in night jungle takeoffs. Early in World War II, vertigo on take-off was the number one cause of lost planes until it was corrected by having one pilot on visual and the co-pilot on instruments from the start of the take-off run.

Airspeed— sixty-five. I put on half flaps and long before the lamp I eased back on the wheel. We were airborne. My eyes constantly scanned the turn and bank, directional gyro, and rate of climb. Everything okay, we were flying directly north to a mountain range five miles out and to our left. At two-hundred feet, I did a gradual left turn; the gyro showed 270 degrees, 260, 250 and finally 175 degrees, our heading to Lethem. Engine sounded better than usual. In ten minutes we were at 4,000 feet, 500 feet above any mountains on our flight path.

Lethem, an outstation along the Brazilian border and 275 miles from Georgetown, was the regional headquarters of the

Commissioner of the Interior, Eric Cassou. Eric was a longtime friend. Years ago, Jan and I had spent many nights as a guest in his house. I flew him around the country, and we enjoyed many a dinner and party at the local Rupununi ranches. It would be good to see him again.

Twenty-five minutes later we were in Cassou's Lethem office. He listened to our story, wrote it into a formal report, and had me sign it. "Don't start a frontier war but don't get yourselves killed either. Diamonds aren't worth it."

We were now on record. He would send a telegram to the Brazilian Consul in Georgetown and to the Governor of the Rio Branco state in Boa Vista. It would explain the situation and danger caused by the Russian immigrant with a Brazilian visa (not for long). The transmission and receipt would take several hours. He called an aide in to start on it immediately. We thanked him and left.

We had no idea what was happening back at the mine site, but as a precaution we didn't circle the airstrip as we normally would. The Mad Russian might well shoot at us. Instead we flew up river below tree-top level. Frank didn't look happy, particularly on the turns where our plane banked at forty degrees and the wing almost brushed the trees. Just before landing, we slow-flighted at water level—a calculated risk. Our airstrip was not long. I had to make this landing with no errors.

From below the bank I couldn't see the camp but neither could Attanasoff see us. If he heard us he was probably scanning the sky. We weren't there.

I came in a few feet off the water with full flaps and heavy on the throttle; a few knots above stall speed. The backwash of our prop over the control surfaces kept us from stalling. As we came around the last bend I saw the end of the strip, kept the wheels just off the water, and at the last moment lifted the nose and cut the power. Frank grabbed his seat and turned to me for assurance that I knew what I was doing and in control. We surged upward just above the runway, stalled and touched down in the first few feet, brakes full on.

Donald Haack

I took in the scene in front of us. Big objects in the runway: huge rolls of light-weight bushes similar to tumble weed. From a distance they appeared solid enough to discourage an unauthorized plane from landing.

An upside down airplane lay twenty feet to the left. Our divers were peering into the cabin. We ran through the tumbleweed, some catching our propeller and making a loud pinging noise. Fortunately, there was no damage.

We shut down the engine, left the plane on the airstrip, and ran back to the accident. It must have happened seconds before we landed. Whoever was in the plane was still inside, possibly hurt, and unable to get out. We approached the pilot's side as the door flew open. Out crawled a strange apparition.

It was a pilot from this area, Bill White, who had the appearance of a science fiction character: his face and upper body had a fluid, dripping movement, as if melting away in front of our eyes.

"Bill, is that you? Are you all right?" He didn't answer. He was unsteady and stumbling. He fell, got up, and fell again. I went over grabbed his arm to steady him. I saw why the melting-face effect: eggs. He was covered in raw eggs. Frank checked the inside of the cockpit.

"Jesus, there's someone else in the front seat and there's eggs and egg shells all over the inside of the plane!"

The opposite door was jammed shut. A voice shouted.

"Get me outta here. . . this things going to explode. Get me out!" He was upside down, hanging in his seat belt and panicking because he apparently couldn't release it. Frank ran around to the pilot's side and peered in.

The person screamed louder, "Get me out!"

Frank reached in and yanked the seatbelt. The screamer dropped on his head. Frank grabbed him by the shirt and pulled him out of the plane. The passenger sat up wiping egg off his face and eyes.

"I think I'm blind. I can't see,"

As he wiped the egg off, his face turned a bright red. Blood and lots of it. His head had struck the control column and opened a nasty wound.

I coaxed the pilot, Bill, to sit down so he wouldn't fall again. We had enough casualties. My attention went to the patient with the head injury. After gently wiping off most of the egg, I exposed the two-inch gash on the hairline. Head wounds, even though less serious, always appeared grim —white flesh and heavy bleeding. This was no exception. It bled enough to get most dirt out, so all it took was to apply pressure with a handkerchief to control the bleeding.

I recognized the voice: Joe Dieter, the ugly American— a more obnoxious person would be hard to find. Bill White was just a harmless bum. I had more encounters with him than I cared to remember—mostly because he tapped into my emergency aviation gas dumps that were stashed around the interior. He borrowed without asking, repaying, or telling me they were gone, something I would find out too late.

This was like a scene out of Keystone Cops. With these two characters in the play, I knew there was skullduggery afoot. The upside down plane was a dripping mess. In a few hours the tropic heat would create an omelet.

Bill White continued his ranting and raving. "My plane, my plane, what am I going to do without my plane? It isn't insured. How can I get back? How can I do business? My plane is everything!"

He sat on a two-foot hummock, pleading as if I were the universal fixer. There were hummock mounds all over this hill, except where we had leveled off the ones to make the airstrip.

Bill struggled to his feet and tried lifting the wing. It didn't move. He fell back, shouting and crying. Meanwhile, Joe kept touching his head, mumbling repeatedly that he was dying and had to get to a hospital right away.

"You have to get me to a hospital immediately. This is an emergency!"

Our divers taking this scene shook their heads. They came over to explain that the local Indians knew Dieter was out to steal our fuel. He talked Bill White into landing on our airstrip after they heard our plane leave. They planned to tell our divers that this was

Dieter's diesel that Mr. Haack had flown in, and it was paid for in advance. They were going to take four 55-gallon drums, roll them into the river and retrieve them downstream.

They didn't expect the airstrip to be blocked, which from appearances it was. They flew in downwind, committed to land in the first few yards. When they saw the large black objects blocking the airstrip, Bill veered off into what he thought was grass but actually deep hummocks. The rest was history. What they didn't know was that the big black objects were only lightweight sagebrush. If they had stayed on the airstrip they would have been okay, but they were sneaking in and the results—disaster and were caught in the act.

Bill switched from hysteria to belligerence. "You ruined my airplane and you're going to have to pay for it." After he stopped shouting, I quietly told him he was caught in an illegal act, and I didn't give a damn about him or his airplane. I could file charges against him for illegally landing on an unauthorized airstrip. That shut him up. But not Joe, who demanded I fly him to Georgetown for medical attention—now.

"This is a life-or-death emergency and you have to fly me out, right now."

"Joe, unfortunately you aren't going to die. If you want to fly down with me next week, I'll take you along for a hundred dollars. If you want me to make a special charter tomorrow morning, it will cost you six-hundred... in advance." I walked away.

Frank spoke up. "To hell, charge the bastard two-thousand dollars or he can rot here— he deserves it. Don't do it for a dollar less."

Though it had been a long, hard, eventful, and stressful day, the anger and tempers were cooling down along with the temperature. At four degrees above the equator, days are unbearably hot, but at night at our 4,000 foot elevation the nights were cool and towards morning, chilly. No one slept without a blanket, but the cooler night temperatures were a nice respite from the unrelenting sunshine and heat of the day.

Donald Haack

At 6 p.m. Frank began his ritual of placing a cloth on our fold-up aluminum table. Johnny Walker Red Label Scotch and a soda dispenser, British style: a bottle wrapped in stainless steel braids with a silver spigot on the top and a small CO_2 cartridge inserted to give it the carbonation. Alongside were two wide-bottom old-fashion glasses on bright red cloth napkins.

You could set your watch by Frank's 6:15 announcement: "The sun's below the yardarm." ...a cue to end the dredging business and enjoy life. Our blue canvas-and-wood fold-up captain's chairs were in place facing the sunset. We didn't have ice. Frank fashioned an ingenious cooling system: a two-gallon jug that slow-dripped water on a towel draped over our drinking water. The dry wind passing over the wet towel created constant evaporation and cooling.

As mad as he was, Frank had some heart. The dastardly duos made a shelter alongside the plane to bed down for the night, and just before dark Frank had our Indian take stew and fresh bread to them for dinner. They sent back a grateful thanks.

This was Frank's moment, carried out to perfection. I had long learned you don't screw up on those procedures. I took our swim and changed into something more dignified: clean khaki shorts and shirts. Bare feet were considered *formal.* We always met the 6 o'clock happy hour deadline. Watching the sun go down was a very short ritual. At the equator there is no twilight as there is in the North. When the sun sets darkness follows in minutes. As the last daylight faded, Frank reached up, unhooked the immaculately polished glass of the already fueled Tillie lamp, pumped in air—instant light. The leisurely drink was followed by a good warm home-cooked meal that put the world in its right perspective.

The bush sounds of leaves rustling, howler monkeys screeching in the distance, tree frogs croaking added to the background hiss of the pressurized Tillie lamp. After the recapitulation of the day's work we took our portable diamond balance, hung on the metal cups on their little chains, put up our wind screen so as not to throw off the delicate balance, corrected for zero balance, and weighed the day's diamond production. Color grading would be

Donald Haack

done in the morning's north light between 6 and 8 a.m., the ideal time for accuracy.

Our formal nightly ritual retained a semblance of civilized living. It was not uncommon for outsiders to go *native* after an extended time in the bush, and we learned from the British who experienced that in Africa and India.

On this particular night it helped defuse the tensions and anger of an exceedingly stressful day, one that started out trying to solve Attanasoff's war but ended up by completely being overshadowed and forgotten by the attempted theft and crash by our "Ugly American compadres."

I mentioned to Frank that with all the fuss we had completely forgotten Attanasoff. Frank was surprised but still focused on the day's accident. He hadn't had his last say on Joe and Bill and didn't want to let them off the hook.

"If Joe wants to charter the plane tomorrow, we charge the bastard fifteen-hundred dollars. After all, they were trying to steal our gas." We compromised on nine-hundred dollars, which was enough to buy four drums of fuel and pay for the round trip. The logic became obscure but that was how it was finally settled.

Surprisingly, in the morning our two interlopers were quite subdued. Bill had decided to stay with his airplane, clean it up, and very humbly asked if it would be okay to set up his lean-to camp and pay us for any rations. He wanted to clean up the omelet before it became unbearable in the tropic heat. He was convinced and obsessed that he could get his plane flying again if we would fly parts in for him. If not he was ruined. He looked pathetic. I had to say yes. We called over the Indians and together we turned the plane right side up. Surprisingly, the damage was minimal, but I would have to take the landing gear down to town and have it welded. He asked me to bring back two tires and fabric to patch the several holes in the plane's skin.

Joe agreed to the nine-hundred dollar fee, which took Frank and me by surprise. We expected a fight or an argument. Joe wanted to leave immediately and said he would pay me in town. I told him his credit was no good. "No pay, no flight."

He stood up, emptied his pockets and swore on his mother's grave that he had no money, just pocket change—probably the truth. I believed him. But no one with dredges leaves the diamond production behind in camp. The owner always walks with the diamonds on his body.

"I'll take diamonds in payment."

There was a long silence. . . a hostile glare that I thought might flare into a physical fight. No one moved. We were all standing. Frank pointedly moved closer to where his pistol hung. The silence intensified. Joe looked at Bill, at Frank, back to me. More scowls. And then suddenly with a little bit of spit coming out of his pursed mouth, he said, "You shit!"

He opened up his belt, reached in, and took out a leather pouch. He sat down at the table. There was the usual haggle that accompanied diamond sorting,

"That's a gem, not an industrial; those are smalls, not sands. . ." the usual grading give-and-take but not open hostilities. The atmosphere was less tense than minutes earlier.

We planned to leave after breakfast. Frank had his *pound of flesh* from Joe and became the perfect host again by inviting them to breakfast with us, which took me completely by surprise. If Frank was being gracious, I could do no less.

They gladly accepted and gave Frank another appreciative audience on his culinary skills He swelled like a peacock to their compliments on such a scrumptious breakfast. Compliments, long and plentiful, were to Frank like water to a plant. He simply needed them to survive. And if one were so obtuse as to not notice how good his cooking was, they were subjected to a long detailed dissertation on culinary art. You only went through that once before you learned that both eating and complimenting were a necessary part of existence in Frank's camp. Even though this wasn't the week end, Frank splurged with eggs—which to my thinking was not a well thought-out choice, corn beef hash, homemade bread, and his famous raisin cinnamon buns, the likes of which I have never experienced before or after. This was followed by copious amounts of freshly ground, hot steaming black coffee that put everyone in a euphoric disposition. Compliments flowed; Frank floated several feet off the ground.

As we finished our last coffee we were interrupted by a commotion on the river. Attanasoff and two of his boys disembarking from their dugout, called up to us, smiling and waving their arms in friendly greeting.

"What the hell—" was all Frank could say. I slipped on my gun belt, waved in return and walked down to greet them. It was just like the first day: all smiles, greeting, talking in Russian, swearing, thanking us for all we did. Frank sidled up to me in his constant slight limp, which was always more noticeable when he was tense.

"Is this the same asshole who was shooting at us two days ago?" he asked so quietly only I could hear it. He never took his eyes off Attanasoff, who in his usual eloquent style took his time to tell why he came: they were breaking camp.

They made "good" diamonds, more than they could spend in a season, the rains would come soon, and they wanted to be on their way for easier traveling. We exchanged slaps on the back, best wishes, and a safe journey. Our departure was delayed an hour until they pushed off in their dugout. Goodbyes were shorter than their entrance, and they soon were back across the river.

After they left, our men told us the real story: The Russian's crew had rebelled. When the shooting started they weren't in sympathy with Attanasoff's aggression, and they drew the line when they saw the *Americanos* shoot back. They quickly realized we could have taken them out and those were warning shots. They told Attanasoff no more diving and to give them their share of the diamonds. They did not want to be buried in the upper Mahu River because of The Mad Russian. They wanted to go home. Attanasoff, with no other choice, agreed to everything.

When they were out of sight, I took off in our Cessna with a bandaged Joe Dieter, whom I left in town. The next day I returned with more fuel. A double crises averted, we settled down to what might be considered a normal three weeks of work before the rains came.

Shooting rapids below camp.

Camp set up on Ireng.

Old miner examines suruku for prospectors. Note dark Ilminite, tantalite indicators in center.

Prospectors examine concentrate. Dugout in background.

12. THE RAINS, MOVING EQUIPMENT

With the Mad Russian situation behind us, we were ready to settle in for a couple of weeks more dredging. That night as we prepared our meal and were sitting down for a drink, Frank pointed to the Mount Roraima area to our north. The sky was dark, but it could have been the Fourth of July in America—light flashes from a celestial camera alternately lit up the looming mountain and then plunged it back into oblivion. The unbroken arcs of lightening from the clouds to the ground produced glowing white dots where they hit the tree canopy. It was a spectacular fireworks show.

This marked the beginning of the rainy season. Heavy rains would accompany wild lightening displays and in a few days the rivers would come raging—in flood mode— to hit us here. There were spots in the river that could be worked in the rainy season, but this was not one of them. Sankar, our head diver, had heard about a big bend in the river twenty miles north of here where I had bought diamonds all year, many years ago.

I remembered there was a small hill in the bend that produced diamonds and was above high water mark. During the rainy season that hill could be worked by hand. The crew wanted to give it a try. They just weren't ready to head for Georgetown and city living yet.

We pulled the dredge ashore and the dis-assembly began. The engine came off first. We removed the pump to fit in the cargo area of the Cessna 180. The canvas roof folded into a small manageable bundle. With six of us working, we made short work of the pontoons by removing the eighteen bolts on each section. We stacked six of the twenty four sections into the plane and filled their insides with tents, tools, cooking utensils, and whatever else we had. It took eight trips in all, and we finished the new camp before dark. Then a small rain squall hit.

Frank had set up two tents, one to cook in and one for our mattresses. He was a miracle man in logistics. And as we were sipping our evening Scotch, I couldn't help but recall what a gargantuan task this would have been years ago. This move that we accomplished in one day, would have taken the better part of a month. Indian *druggers* would require three to four weeks to transport the equipment. While they were doing that, a half dozen more Indians would be needed to find, cut, and hollow out fifty-foot trees for the pontoons. We drank a toast to progress.

In two days, we had the dredge up and running. It was the first time any mechanical equipment had ever been in this section of the river, and it yielded a tantalizing amount of diamonds. Sankar and Adolpho were prospecting in the morning when the first batch of diamonds came up. They didn't stop for lunch, and Frank had to convince them to slow down and eat something or they would pass out from exhaustion. They reluctantly took short breaks, but every check in the sluice box produced diamonds, and the incentive for uninterrupted work was strong.

Everything came to a head in the evening. Instead of shutting down at sunset, Sankar and Adolpho were still working the lucrative diamond pocket. Both divers were convinced that the biggest deposit was still there in a long section of gravel. They wanted to continue. Frank, the perfectionist, wanted everything on schedule. The argument became nasty. I had to step in and make the decision. Frank insisted that they could start early in the morning, and that it was too dangerous working the dredge at night with only Tillie Lamps for light. According to Frank, everyone was exhausted, and we were an accident waiting to happen. He wouldn't budge. I talked to Sankar and Adolpho—in order to keep peace in the camp, we would shut down and resume early in the morning.

At the break of dawn everyone was up and eager to move out. We were having coffee when one of the men made a comment about the river noise becoming louder. We rushed down to the water's edge. The dredge was swinging wildly from side to side. It was firmly secured but the rushing water bounced it around like a cork. What was a calm and complacent river the night before, was now a raging torrent. There was no way a diver could get down into that maelstrom to guide the intake nozzle on the dredge. We had to wait. The surge lasted two days before the men could suit up and check the work site.

Sankar came up after a short fifteen-minute dive. He tore off his mask and threw it on the deck. "There isn't a fucking bit of gravel left down there. It's been wiped clean, every bit of it—just plain bedrock. Once the gravel was disturbed, the river washed the rest of it away." He swept his hand in a broad sweep of the river. "Somewhere."

Some hot and heavy discussions followed. Sankar and Adolpho wanted their share of diamonds. They were breaking camp. They were mad at Frank and would have nothing further to do with him.

"Keep your goddam food. We came here for diamonds. We don't know why you're here, but it's for sure not the same reason we are."

And they left. Without divers, we decided to pack up the equipment onto the bank and close down for the rainy season.

I couldn't help but think that the quest for diamonds was a constant challenge. What God had hidden was the first obstacle. Man's ability to extract them was another. And the other factors—weather, sickness, and self-made rules all added up to making the challenge more difficult and in many cases impossible.

Donald Haack

13. HAZARDS OTHER THAN FLYING

Bertie and I agreed that beside our most onerous task of transporting bodies from the interior we had to run the gamut of other hazards that tested us on the twenty-five mile Atkinson Road, the only link between Georgetown and the international airport.

Un-paved with deep and numerous ruts, the road was one continuous obstacle course; a walking path, bike riding lane, animal path of sheep, goats, horses, cattle and most of all chickens, all of which had the free run of the road. The bigger cars and trucks carrying international cargo, bulk sugar cane, fresh fruits and vegetables grown in the vast fields along the road commanded the highest priority because of their size. To describe it as chaotic and stressful would be a mild interpretation. This unavoidable twenty-five mile stretch also had the distinction of producing some of the most awful smelling putrefied gasses of fermenting sugar cane from the four factories adjacent to the road.

For visiting businessman or tourists, this colorful experience was the first impression of British Guiana. For most, if given a choice, it would be their last.

Bertie, my driver, and I were challenged by that road twice a day and mostly in the dark. To complete two interior flights a day, the first had to be at the break of dawn. The second one ended only minutes before dark, which meant most road trips were in the dark. Traffic did not decrease at night. It was only harder to see. Some drivers perceived that as an advantage—you couldn't see what you hit.

One Saturday when I had only one flight, Bertie and I drove up in the daylight. I couldn't help but admire Bertie's excellent driving but it was almost like a game with priorities; avoid trucks and cars first, then pedestrians and bike riders, followed by animals according to size. Chickens didn't have a rating and had the highest mortality rate. There were so many it apparently did not make much difference, at least not enough to keep them fenced or caged. And they did provide instant road-kill for the impoverished. The road ruts were somewhere in the middle range. If you ignored those, you wouldn't have a car or truck to continue the game.

I appreciated Bertie's driving ability to avoid countless obstacles continually bombarding our car. Then I saw it. Venting his anger, an East Indian man struck a hard blow on a large mule walking along the road. He reared up, leapt forward on his hind feet and landed in our path. The impact from our car threw him up in the air and as if it were happening in slow motion, the mule slowly spun in a semi-circle, hovering a couple of feet in front and above the car. That is, until gravity once again took over. We were doing a modest thirty miles an hour when the mule descended onto our bonnet (hood), collapsing it and shattering our windshield. The mule slid off onto the road. We were certain we killed it. It raised its head, shook out the cobwebs, got up and walked away. The car was in shambles. It wasn't going anywhere.

I don't know if I was lucky in having a driver as qualified as Bertie or we were in some way protected. For the most part, we escaped any serious injuries or killing any of the local population.

There was one incident that shook up both of us. We were driving in the dark, engulfed in a blinding rain coming down in sheets—forcing us to creep along. Staying on the road was a major challenge. Oncoming cars slowed to a near stop in order to pass and not slide off into the ditch. We passed a car coming in the opposite direction. Our windscreen wipers on full speed were not keeping up with the deluge of rain. After the bright lights passed, we struggled to get our night vision back. We hit a big bump. I thought we went off the road.

"No, Cap, we're still on the road. There must've been something on our side of the road when we passed the car. Bad bump but I guess we're all right. Turned out later that we were okay but the *bump* was not. That night when we returned to the ramp for refueling, we heard the whole grisly story.

It was a slow night. All international traffic was cancelled. George, the Shell man who serviced our plane, was bored and had a few drinks too many. His boss told him to go home and sleep it off. His car wouldn't start, the phone system was down from the storm, so he decided to walk home down the Atkinson road—in the dark. They finally found him, or what was left. He apparently was struck by a car or truck and thrown across the road where for the next six hours, cars and trucks continued to run over him without knowing what it was they were running over. We were one of them. It was terrible and afterwards for weeks Bertie and I harbored guilt and depressed feelings over the incident.

One of the reoccurring nightmares I had was of landing on the interior airstrips with crowds of people running, converging to greet the plane and being first for a seat to town or snatching the mail bag for the trade store. The rushing miners were oblivious to the airplane's prop that continued to rotate for a few seconds after the plane stopped. I tried every trick; turning the plane 180 degrees from the crowd as I shut down the engine, opening the door and frantically waving them away from the front of the plane, flashing the landing lights if it were dusk or cloudy or any other trick to keep the madding crowds away from the lethal propeller. It was only a matter of five seconds between stopping, shutting down the engine and waiting for the prop to cease turning.

One morning I landed in Imbambadai where there already were people lining up along the strip. I landed further down than usual so I wouldn't be in the middle of the waiting line when the plane stopped rolling. I turned off the strip and shut down. To my surprise, even though most of the crowd was behind and on my right, one man carrying a knapsack had sprinted around the plane and approached from the left rear. He passed a few feet from my cockpit heading forward and I knew before it happened he would run into the turning prop. He never saw it and I heard the metallic *clang* as he stopped the prop and was plunged into the ground. My worst fears became a reality and I didn't know what to expect.

I quickly exited the right hand door, pushed aside the onlookers now on the scene and bent over to examine the prone victim. He was face down, the knapsack still on his back, but a steady line of blood was coming out of his shoulder. My first impression was that he was dead. I thought the prop hit him in the head but on closer examination, I saw it had lacerated his ear and struck his shoulder, opening a big gash from the shoulder down the arm. Bone was exposed and the bleeding was profuse but was not arterial. It wasn't pulsing and I knew from my limited anatomy courses that the artery was on the inside of the arm. He moved and groaned. I ran back, got my emergency medical kit and took out the heavy gauze bandages. One of the women standing alongside appeared to be more than just a gawker.

"Can you help?" I asked. She nodded and I pulled her hand down with me as I packed the open gash with the gauze. I wrapped it around as tightly as I could, placed her hands on it and told her to apply pressure—it would stop the heavy bleeding. She came through like a trooper. I recognized her as one of the two local prostitutes, a tough lady. Then we eased our patient onto his back. His eyes were open, but glazed over.

"Can you hear me?" He nodded yes. "You know what happened?" Again a nod yes. "You are lucky you aren't dead, you know. You're going to be all right. Can you swallow a pill with some water? It will keep the pain down. You probably don't feel much now but you will in a while. I'm taking you right to town and they'll have an ambulance take you to the hospital. You'll be okay." He seemed to understand. I took out a morphine pill, poured some water in a paper cup, and held it to his lips. "Can you take this?" He started to raise his other hand to hold the cup

Donald Haack

and it put him off balance and we had to readjust the bandages to keep them in place. "No, no, I'll hold it. You just drink. Here's the pill and swallow it down with the water. You'll feel better in a few minutes." He was going into shock already.

I'm going to need a couple of you strong guys to lift him gently into the plane as soon as I straighten out the back area to let him stretch out." I asked my lady friend holding the bandages, "Were you going to town?"

"Yes, I hoped to. Been looking for a ride for a couple of days."

"Okay, you're lucky. You not only have a ride, but a free one. Can you keep holding that bandage on him on the way to town? I've got more here that I'll give you. When they pick him up, walk with him and keep pressure on the bandages. He's going to be all right but that cut is right down to the bone. Don't know if the shoulder bone is broken so we'll secure it before we lift him. You okay with that?" She nodded yes.

We eased him into the plane, propped up his feet, put several blankets around him to keep him warm while the morphine took effect. It was all we could do for him to prevent his sliding into shock. It worked.

They had an ambulance waiting at Ogle Field and the medics took over from there. I checked a couple days later to find that no bone was broken but his arm had taken a nasty slice. They managed to stitch and tie together most of the muscle and ligaments, and sewed him up without complications or infections setting in. In a couple of weeks he was up and around and eventually regained almost normal use of the arm. It could have been a lot worse.

14. TRAGEDY ON THE IRENG

The first two safaris of the second season went without a hitch. We expected the good luck to continue. It didn't.

We set up camp on the border along the Ireng River for the eight men in our third safari. The river produced a small but steady flow of little diamonds, and in the first few days everyone had already surpassed their *guarantee* of a half carat of diamonds. After finishing our drinks and settling down for dinner, we were hailed by two people coming down the path. We recognized them immediately and asked them to join us for dinner. They eagerly accepted. Wes and Bill, who were on an earlier safari, had decided they wanted to stay permanently on their own. Both were very likeable and our crew had a great camaraderie with them. They enthusiastically gave an account in detail how they were doing on their new adventure.

Wes, in his late thirties, was in the insurance business. He joined with Bill just out of high school and had worked a few months at Globe Union Corporation. Bill had lived with his parents in a modest part of Milwaukee. His biggest assets were his unending enthusiasm and willingness to pitch in to do more than his share of work. He was a joy to have around. Wes had walked away from a lucrative insurance agency in Shaker Heights, an upper scale suburb of Cleveland. We heard later he'd separated from his wife, wanted a change, and was looking for something more out of life. The fact that Wes and Bill were exact opposites became the reason for a strong bond between them—and the basis for their joint diamond mining venture. They asked me for help, and I was more than happy to steer them in the right direction. That largess extended from everyone in the present safari and was manifest in our group's quick and positive response to their request for help.

They purchased a dredge in King's village, downstream from us. Don Lynn, a miner and oil wildcatter had a business in Sierra Leone, Africa, that was going down the tube and demanded his immediate attention. Don sold out to Wes and Bill. It was a win-win deal for both of them. Bill and Wes managed to maneuver the dredge upstream, close to the area they wanted to start dredging but needed more manpower to pull the rig through the last section of swift water. It was too big to maneuver by themselves and one Amerindian. They asked if we might help them on Sunday. The crew enthusiastically and unanimously approved. It would be a work/fun outing and Frank would pack the food and beer for lunch.

Sunday, Frank gave us a bright and early start with his gourmet breakfast of pancakes, bacon, hot coffee and a fruit compote: packaged dried fruit simmered in cream with a dollop of Demerara Rum for flavor. It was dubbed Frank's *Ireng Fruit,* a true delicacy.

After breakfast, we hiked the trail to where Bill told us they secured the dredge. We brought extra ropes. Bill and Wes were waiting for us with their small boat on our side of the river. Bill yelled and waved for us to come to where the boat was tied. There were six of us, including Ed, Chuck, and Lon and were ferried three at a time. We were in a big river pool but up ahead we saw their challenge. The river narrowed between two sets of rock outcroppings, forcing a swift channel of water to gush through. Above that, the river spread out again. My thoughts were that we should be able to pull the barge through that channel and be finished in less than an hour. We tied on our extra-long rope so that two sets of men could pull at the same time.

Bill and I started out on the barge with Domingo, and since Bill was a non-swimmer, it seemed safer than on the bank where he might fall in. He wore heavy leather boots, not well suited for the bush. In the tropics, constantly wet feet are guaranteed to result in a fungus infection. We all wore canvas Keds—they dried out fast, were lightweight and cool.

We reached the end of the pool near the fast water in a matter of minutes, and then planned our next move. I asked Bill to go on shore and gave him my camera. The other men pulled the ropes from along the edge of the rocks while Domingo and I handled the barge. The men on the bank began the pull upstream. Sitting on the side of the barge and using our feet to push off from the rocks, Domingo and I kept the barge from hanging up on the protruding rocks. Progress was slow in the fast current.

We were halfway through the swiftest current when Ed called. They were moving further upstream onto another location where they could pull us all the way through. In the meantime, Domingo and I kept the barge from scraping and hitting the rocks. It wasn't an easy task. Chuck and Lon were in knee-deep water, holding the barge in place with one rope, while the other three climbed on the rocks further upstream. Bill was on the bank, safe and dry, or so I thought. Moments later I heard shouting.

Bill, always willing to do more than his share, decided to help pull. He jumped onto the outside rock for more leverage. He was about to pull on the slackened rope when it went taut and hurled him into the water. Wes, standing alongside, lost his balance and fell in behind Bill, who disappeared beneath the waves. Wes floated down the channel. I picked up the rope that lay on the pontoon, shouted, and threw it in front of him—no response. He stared straight ahead, his head low in the water, as he passed by our barge— too far out to grab him. He was a good swimmer so Domingo and I waited for Bill to surface. When he didn't surface immediately, we both dived in. Nothing. The barge was loose and going downstream with us alongside. We climbed back on for a better view, hoping we would see Bill surface at least once. He didn't. I glanced down river to see how Wes was faring. Domingo was already pointing. Wes's head was lower in the water, and he hadn't made the first move to swim ashore. Then suddenly as we were watching, he disappeared. We both dived in and swam as fast as we could towards the last place we saw him.

This didn't make sense. Wes could swim. He didn't hit any rocks in the center of the fast-moving water, and now, midstream in the big calm pool, he simply slipped under water. When we reached the place we last saw Wes, I told Domingo to dive upstream. I would criss-cross the area further down.

Minutes slipped by. No results. Domingo shouted. I joined him. He was exhausted. I was, too. He pointed down. I hyperventilated and dived down. At first nothing, then I spotted a white form— Wes's undershirt. I continued down. My ears popped and hurt. I was running out of breath, but I felt I didn't have enough strength to repeat this again, and we may not find him on a second dive. I went down as fast as I could, grabbed his shirt, and struggled up towards the surface. I was completely out of breath and exhausted. Domingo appeared on my right. I grabbed his hand and put it on Wes's shirt. I needed air—and fast. I shot up the last few feet and gulped in fresh air, hyperventilated quickly and submerged again. Fortunately, Domingo was close to surfacing. He grabbed his throat, indicating he, too, needed air. He let go as I grabbed Wes. We broke the surface together. I tried holding Wes's head out of the water, but he was dead weight. Neither Domingo nor I had much strength left.

Pulling Wes, we swam for shore, a mud bank with small scrubby trees clustered together. The trees gave us handholds in the slippery mud, but the foliage was so thick it impeded us from pulling and pushing Wes onto the bank. We finally maneuvered him through a narrow slot of small trees. I turned him over and held him by the waist to allow water to come out. None. I placed him on his back and gave the standard CPR procedure: mouth-to-mouth breathing alternating with pumping the chest to activate the heart to keep blood circulating. I quietly gave thanks I took CPR courses. Even though this was the first time I actually used it on a patient, the procedure came back easily. I lost track of time. I kept up the mouth-to-mouth until I blacked out from forced breathing and the exhaustion from diving and dragging. Domingo tried to stop me several times.

"Too late. Too late. He gone. He dead."

I wouldn't listen and continued between the millisecond blackouts. I wouldn't give up. Domingo persisted and tried to pull me off, to no avail. Then as I was about to give mouth-to-mouth again, I glanced down. A fly crawled out of Wes's nose and flew off. The significance brought me back to reality. Wes was dead. I scrambled to the path where I sat and broke down. I didn't want Domingo to see me in this state. In a few minutes, two of our men came by. They already knew what happened. They asked if we were able to retrieve the body. I pointed to the river.

"On the bank. It's Wes. Domingo's with him. Can you help bring him out? I can't. Bill?" I questioned. Ed shook his head no. He followed my path to the river and returned within minutes.

"I'm going back where we crossed. There's a tarp there that we'll need to carry him back to camp," Ed explained. He left while an Indian cut two small trees and trimmed off the branches. Chuck and the other Indian disappeared. They pulled up Wes's body onto the path the same time Ed returned with the orange tarp. In minutes they had Wes in the tarp with both poles secured into a sling. The four of them each took an end of a pole, placed it on his shoulder, and we took off on our way back to camp.

I was physically and mentally beyond exhaustion and barely kept up on the trek back. I collapsed in my hammock. A couple of hours later Ed and Chuck brought me back for dinner, but not without plying me with drinks first—their answer to alleviating my semi-state of shock.

We reviewed the scenario over and over, trying to figure what went wrong. Lon and Ed said Bill unexpectedly jumped on the rocks, grabbed the rope before anyone could stop him, and moments later he was in the water. He went straight down and never re-surfaced. We discussed Bill's inability to swim.

When I first met Bill in Milwaukee last November he admitted he couldn't swim, I suggested he go to the Brookfield Diving

Center for swimming lessons. I don't know if he did, but both Lon and Ed said Bill couldn't handle himself at all in the water. With those heavy boots on, he went down like a lead weight. But Wes was the enigma—an excellent swimmer. We were all puzzled as to why he drowned. It just didn't make any sense at all.

The next morning I flew Wes's body to town and notified the American Consulate about the accident. They were a big help. They said this sort of thing happens regularly in the bush, as it does in construction work, but the project must go on. They urged me not to cancel the next set of tourists coming in the following week. They would notify the families of Wes and Bill while I cabled Jan. I sent a Western Union telegram, the quickest way to communicate.

1965 Feb 6th AM Elm Grove WIS
BILL AND WES SLIPPED OFF ROCKS DROWNED IN IRENG YESTERDAY FOURTH BILL OVEREAGER NOT SWIMMER WES PUZZLE CONSULATE AUTHORITIES HELPFUL THRU BLACK 24 HOURS AFTER TRAGEDY STATE DEPARTMENT HANDLING DETAILS AND NOTIFYING FAMILIES I WILL WRITE EVERYONE SAYS CONTINUE KEEP BUSY SO WILL MEET GROUP AS PLANNED LON CHUCK ED FRANK OK REASSURE THEIR FAMILIES IF NECESSARY LETTER FOLLOWS LOVE DON

Later in the afternoon I received a telegram from Jan.

6 Feb 1965 DIASAF GEORGETOWN
BILLS FATHER CALLED HOLDS NO ANIMOSITY IS CONSIDERING FLYING DOWN CAN YOU PHONE HIM FROM GEORGETOWN OR GIVE HIM HAM CALL LETTERS FOR HIM TO CONTACT YOU NEEDS DESPERATELY TALK WITH YOU WANT DETAILS INFORM ME BY RETURN CABLE IF POSSIBLE I AM WITH YOU IN SPIRIT CHIN UP AND DO BEST FOR OTHERS WILL HELP ALL I CAN LOVE YOU JAN

That evening I sent another telegram.

RETURNING TOMORROW FOR BILLS REMAINS IF
RECOVERED DETAILS TOMORROW OR MONDAY WILL
CONTACT BILLS FATHER EARLIEST THANKS FOR HELP
LOVE DON

1965 Feb 8th Jan Haack Elm Grove Wis
RECOVERED BILL YESTERDAY DIFFICULT PHONING
WRITING FATHER DETAILS DON

Two days later I picked up four more safari clients. I explained
the accident as well as I could. I was surprised at their reaction. It
was almost as if it were an added stimulant to the adventure with
the life-threatening dangers in the weeks ahead. It may have been
for them, but it wasn't for me.

Wes's brother and estranged wife notified the Consul that they
wanted a complete investigation. The fact that a good swimmer
could drown wasn't acceptable to them. But then again, it wasn't to
us, either, so I couldn't fault them in their thinking or suspicions.

The following day I received six summons: four for our
tourists and two for Frank and me. We were to report to the
Commissioner's office in Lethem for a court hearing. It meant I
had to leave our four new tourists, but they were in good hands
with Domingo and three other Amerindians. They assured us they
would be just fine.

It was my first taste of English law courts. Each of us
individually had to stand in front of the court, behind a podium,
which we were not allowed to touch. We had to keep our hands by
our sides at all times during the questioning. Standing immobile
with hands at our sides was more difficult than answering some
of their totally irrelevant questions—and then specific ones like,
*How could there be damage on Wes's body if he was in clear water
at all times?* At first that had me stumped too, until I realized that
when we placed him in a tarp slung over two poles, we had to
walk through a bush trail for over an hour. It was a hard ordeal and

in some of the narrower areas, the carriers stumbled and dropped Wes. The body swung back and forth and in many places bumped against rock outcroppings and trees on the narrow path.

Each one of us had to testify separately while the others were confined individually in small rooms. The court wanted to see if our testimonies matched and that we weren't making up a story. After a full day of the "third degree" questioning and being admonished every few minutes about standing up straight behind the podium, keeping our hands at our sides, not brushing away the sweat from our brows, we were told they didn't feel we committed any heinous crime. Several days later they advised me that the autopsy showed Wes did not drown, which answered my question why no water came out of his lungs when I bent him over on the bank. Wes suffered a massive heart attack, a heart condition we were not told about until after the inquisition. We also learned that Wes's family knew about his weak heart.

The shock of seeing Bill, who couldn't swim, fall into the swift channel apparently triggered Wes's heart attack. It didn't make the situation any easier, but it at least answered the question that plagued us.

We finished off the rest of the season without any more incidents. I don't think we could have handled anything else.

I kept thinking, *the earth has diamonds but doesn't give them up easily.* It extracts a high cost.

Don, Harry, Domingo navigating rapids.

15. EXPANDING HORIZONS

L osing Wes and Bill took its toll on me, even after a couple of months passed. Though they had gone on their own and were not a part of our safari, I'd established a close relationship with them and felt the loss keenly. I struggled through several more safaris and put on a good show resulting in some very happy campers. In between safaris, I flew charters that brought in more income than the safaris themselves, a fact Jan pointed out on several occasions.

Jan and I were not happy living apart for long periods. That was never a part of our plan. We both wanted a change for the better. When the safaris started, we didn't have much choice. There was no income coming in, and we had an opportunity to take advantage of a situation that macho men wanted to do and didn't mind paying a premium price for the privilege—our unique *Diamond Safari*.

Reviewing the financial statements, we realized that the greater portion of income came from charter flying. There was a definite need. We had the right idea in bringing the Seabee amphibious plane down, but we didn't anticipate that our pilot, Jimmy, would go money-berserk and kill himself. The answer seemed clear— buy our own seaplane and hire a pilot. With one land plane and one seaplane, we could cover any area in Guiana, and it would be a good start for an expanded and much needed air service that the government was encouraging.

The Minister of Communications, Mr. Correia, had several dredges and trading stores in the interior and told me on several occasions that when I serviced him and his friends with our dependable interior flights, his operations did very well financially.

But Georgetown was not an ideal place to bring the family to live. However, the beautiful Caribbean islands were a couple of hundred miles to the north. So I wrote to the tourist boards of the closest two, Trinidad and Grenada. When I had flown over on previous trips, Grenada appeared to be the prettiest. I immediately had a reply from Trinidad with a pack of brochures and annual rentals. I had to write a second time to get a response from the Grenada Tourist Board, my first choice. Unlike Trinidad, which was industrialized with oil, refineries, ship-building, and manufacturing, Grenada was an idyllic tropical island, still untouched by civilization, with the charming town of St. George surrounding the Carenage where ships and sailboats abounded.

I was sorely disappointed by the response. Grenada's rentals were several thousand U.S dollars a month, much higher than the best areas in the States. I shelved the idea. After the last Safari group left, I brought my Cessna 180 into Ogle Airport for a full inspection and servicing by Harry Wendt, a friend and head engineer of B.G. Airways maintenance. The round trip flight from South America to Milwaukee was over 4,000 miles and I wanted the plane to be in first class condition. Half of the trip would be over the Caribbean Sea.

Three days before I left, Chuck Farmer, an American and a diamond diver, came to me in the Hotel Tower Bar. "Man, I've looked all over for you. Heard you were flying to the States and I was hoping you hadn't left yet. Nobody knew where you were. Had a lean year and funds are low. Any chance of flying up with you? Be glad to share the gas and other costs. I don't have enough cash for a commercial airline ticket."

Chuck was one of many of the diamond fortune-seekers to come to Guiana. Many came, few left rich, but there was always that chance of the *big strike* that kept them coming. Chuck was an

A & E airplane mechanic who made a good salary in the States. And he, too, was drawn down here with the hope of striking it rich. It hadn't happened in three years.

"Chuck, I'll tell you what...if you can give Harry Wendt a hand with the plane's inspection tomorrow and Wednesday, you can sit 'right seat' and forget about paying the gas. When you next hit it big time you can pay. Besides, if you're going to fly all the way to Florida with me, I know you'll do a perfect job on the inspection. Deal?"

He grinned and nodded yes. Chuck Farmer was the easiest-going of all the Americans who came down. You couldn't help but like him. His flying right seat would make the very long trip go fast.

"Thanks, I really appreciate it. I'll go see Harry Wendt right now." And he was off.

The rainy season had officially begun, but the day of our departure was beautiful and clear. We left in the afternoon, later than I planned, but we could easily overnight in Trinidad, Grenada, or St Vincent. The latter two would be a first-time visit. It might be fun to overnight in Grenada to find out why the place was so expensive. But then, we might get hit with a high-priced hotel. *No, Chuck and I are both bushmen and in a worst case scenario, we could bunk almost anywhere and hang a hammock. We didn't need the amenities of a high-priced tourist hotel.*

We landed at Pearls Airport on the east side of the island, tied down the plane, and went through the usual rigmarole of flight manifests, customs, and immigration. We spread the word around the airport that we wanted to be near St George but didn't want an expensive hotel. Was that a possibility? Our short stocky female immigration agent at first didn't answer, stamped our passport, and wrote something on a piece of paper that she stuck in the passport.

"You'll find these accommodations at the Ross Point Inn quite satisfactory, the food excellent and the owner, Curtis Hopkin, a very charming Grenadian. I'm sure you will enjoy your stay, and it's not expensive." Then finally a smile. "Welcome to Grenada."

At that moment, I felt welcomed and wished I could make this island my home. I had seen enough from the air to want to explore more.

If there were ever a *perfect tropical island*, Grenada would be first in line. The taxi took us over the lush rain forest mountain of Grand Etang with a lake at the top of what used to be an active volcano. Coming down the other side, we had glimpses of the western shore, the deep blue Caribbean, and a few views of the town of St. George, semi-circling the *Carenage*. It was picture-perfect and got better as we approached. We passed hundreds of small painted houses with well-kept gardens, coconut palm trees, bananas, cashew, and mango trees. This island obviously had a fertile soil with ample rainfall.

We arrived in the southern part of town and drove in front of the quaint buildings along the bay. Our driver explained that the working boats were loading cargo to take to the islands, and the sailboats were anchored while their crews sat in the second floor of the Nutmeg Restaurant, overlooking the bay, enjoying rum punch for "tea."

The word *Carenage* was derived from the French word *careenage,* which was to tip or incline a vessel for cleaning, repairing, or caulking while turned on its side. During a two-hundred year period, the French and the English alternately took possession of the island, hence the many French names. Some of the local boats still used the protected shallow bay for careening their boats. They couldn't afford the costly charge of the Grenada Yacht Services lift, which raised boats out of the water for repairs. GYS had a long waiting list with high-priced charter boats and private yachts.

A few minutes later we pulled into the picturesque *Ross Point Inn,* a prominent point overlooking the blue Caribbean, the Grand Anse sugary white sand beach to the south and St. Georges to the north. *Could there be a more scenic setting?*

A young, nicely dressed black boy opened the door and greeted us with an ear-to-ear grin and a mouth of sparkling white teeth. "Welcome to Ross Point Inn. You will enjoy your stay here."

Donald Haack

He took our bags and motioned us to follow. The immigration lady called ahead, so Ross Point was expecting us. As we passed through the main dining room, a young lady intercepted us with a cool, refreshing rum punch and another *welcome to Ross Point Inn* accompanied by a correspondingly welcome smile. If Chuck and I had any stress and tension leaving Guiana and flying this leg, we lost it like shedding heavy clothes. We felt we were in a tropical heaven.

"Dinner will be served in half an hour, which should give you time to freshen up. Would you like me to bring you another rum punch to your room?"

The couple of sips of the punch warned me there was an ample serving of local rum. I could feel it already.

"No, thank you. It's delicious, but we'll wait until we have dinner. We might have another, then." I spoke for both of us. I knew Chuck could handle his drinks, but I wanted to make a good impression and that meant we should stay sober if I intended to find out more about this island.

After showering and changing into something more suited to the island, we were presented with a six-course dinner of local foods and home-style cooking, a fixed menu. We didn't have to make choices, but I couldn't have done better with a five-page menu to select from. The food was out of this world. As we finished, a tall good-looking man came to our table and asked if he could join us. He introduced himself as Curtis Hopkin, the owner of Ross Point Inn. As the immigration lady said, he was charming and had a wealth of information about the island, its history, and accommodations. I related the exchange I had with the Grenada Tourist Board and how disappointed I was that I couldn't afford to live here. And what exactly made it so expensive?

He tilted back in his chair, threw up his hands and laughed uproariously. "Our Grenada Tourist Board is extremely efficient at turning away tourists…unbelievable. Tell me what you have in mind. How many are there in your family? Postpone any plans you had on taking off early tomorrow. I will take you out to L'Anse Aux Epines on the south of the island and introduce you to someone who has cottages for rent…and not for thousands of U.S. dollars. How about $400 BWI dollars, which equals $240 your money for a three-bedroom house overlooking the bay? Then I'm going to take you to the Westmoreland School where you can meet the headmistress and decided if you want to enroll your children. Breakfast is at 8:00 a.m., and I'll pick you up at 9:00. You're in for a pleasant surprise. Good evening, have a good sleep. There are no mosquitoes or bugs, so leave your windows open and turn the overhead fan on medium. We don't have air conditioning. With the trade winds, we don't need it. Some of the fancier tourist hotels have it but not here at Ross Point Inn." He said it with pride rather than as an excuse.

We retired early and had the best night's sleep I could recall. We were up at daybreak and enjoyed the coffee placed on our veranda. Chuck and I exchanged glances as we sipped our coffee and gazed at the clear blue ocean below.

"Sure beats Guiana with its humidity and muddy beaches, doesn't it?"

The breakfast was the best I had had since I left the States two months ago. Curtis came to pick us up promptly at 9:00. We drove to the south end of the island where we met Christine Braithwaite, who with her husband owned a considerable acreage of the peninsula. Christine couldn't have been more friendly, fun, and helpful. She showed us her mother's, Mrs. Monroe's, cottage, which appeared to be the best suited for our family. I was so taken by it, I was ready to put down the deposit and first month's rent.

"Hold on, hold on," Curtis interjected. "I promised to show you the school and headmistress for your children first. Christine's children are in there also. What do you say, Christine…about Westmoreland School?"

"Curtis, you already know. We all love Beryl Ball, our headmistress. She does a fabulous job. We're all working hard to figure out how we can build a school and get the children out of the makeshift two-story house they're in now. That's the only drawback. Show him the school, Curtis. Maybe he won't want his children going to a house rather than a school building. Off with you now." Christine directed her attention to me. "I'll be here later if you still want the house. We'd love to have you as neighbors." I knew she was sincere, and I wanted to be their neighbor also.

We went into town to the top of one of the hills where Curtis pointed to an old house. Several well-dressed children were going in and out. We met the head mistress, Beryl Ball. Like Christine, we took an instant liking to each other. *Jan, too, would love this woman.* She was British and had lived in South Africa. She was a no-nonsense person when it came to education. It was the British system, which Beryl assured us was as good, if not better, than the American system. I enrolled three of our four children, went back to Christine Braithwaite, and gave her a deposit.

Never had I been to a place that seemed so much like home—as if I were actually coming home, not arriving at a new location. Grenada felt as if it had been waiting for my family and me for a very long time…and we finally showed up. I wondered what Jan's response would be when I told her we had a house in Grenada ready to move into and our children were enrolled into the fall semester of Westmoreland School?"

As I suspected then and time would later prove, Christine, Beryl, and Jan became close friends, an association that would last a lifetime.

We thanked Curtis profusely for not only the great hospitality, good food, the background on the island, but the beyond-the-service of finding us a house and making sure our children were taken care of in school. The outreach of the Grenadians was an incomparable experience in itself. It was indeed, a *coming home.*

We took our time, enjoyed the leisurely pace of Grenada, stayed the night, and got an early start in the morning. We refueled in San Juan and, with a strong tailwind expected, our ETA in Ft. Lauderdale would be before dark. Clearance in San Juan was faster than usual. After we filed our flight plan and checked the weather in the Flight Information room, we were asked to come into the back office. I didn't know what to expect.

The man in charge brought us to a large wall map and introduced himself as head of the FAA in San Juan.

"We received word that a small plane that left here this morning did not report over South Caicos and is two hours overdue in Nassau—that's an hour past the time their fuel would have run out. We've contacted the islands along the way. Nothing showed up. Do you think you might fly their proposed route? I have it here…and do a visual check? I know that's asking a lot because you need to fly at 2,000 feet instead of your proposed 11,000 where you can lean-out fuel and catch a better tail-wind. We have no other scheduled planes in this area that can do an enroute search and rescue. It's a blue-and-white Cessna 172. They have life jackets and an approved life raft on board. They also have a hand-held emergency signal transmitter, but none of the airlines passing at 30,000 feet have picked up any signal.

Whether you're a flying the bush or over the Caribbean, a request for doing *search and rescue* for a fellow pilot is not an option—you do it. We took a copy of the flight plan along with the radio frequencies to monitor and departed.

Flying at 10,000 feet above the light fluffy clouds is a smooth ride, but down at 1,500 to 2,000 feet the air is turbulent—a hard three-hour flight. Chuck scanned the sea outside his window. I did the same from mine while I kept the plane on autopilot. Every few minutes we did a frontal sweep, then back to our side windows again. We switched radio frequencies and called on each one in the hopes of getting a response. Nothing. Flying at low altitude used up our precious fuel and necessitated a stop in Nassau, where we contacted their ARB, Air Registration Board, and gave them the negative findings to be passed on to San Juan. They were aware of the overdue plane. There was no news on it. Not a good sign.

We were fatigued after flying bumpy weather for three hours. We reached Ft. Lauderdale after dark. On the landing, the plane first swerved to the right and then to the left. I pulled off the runway at the first turn-off and advised ground control I wanted to check the aircraft. Something was wrong: our tail-wheel tire had blown out on landing, and we were trailing a half-torn tire that skewed the plane back and forth. Air Traffic Control guided me to Customs and Immigration and had someone from the maintenance hangar come to see how they could help us.

While we cleared Customs and Immigration, the maintenance man took off in his jeep to find us a tire. An hour-and-a-half later we had a new tail-wheel tire and taxied out of the Customs area. I dropped off Chuck and since the weather report was good, I decided to fly a couple of hours so I wouldn't have a long trip the next day. I landed in southern Ohio, but not before I realized that I did not clear the 7 ½ carat diamond that I arranged to carry along with the proper export papers from Guiana.

While Chuck had worked on the plane in Guiana, I'd arranged to ship a large packet of diamonds through the usual bank channels. To save time I wanted to carry the big diamond with me to cut and polish it in Chicago, possibly for publicity for the safari or to sell it to help our cash flow. Shipping through bank channels always took weeks, and I didn't want to wait. In the calm of flying back, the horrible thought came up that I hadn't declared it on my customs declaration. The harried day's search and rescue flight along with the flat tire incident totally obliterated the thought of the diamond.

I got on my radio and called back to Ft. Lauderdale and asked to speak to customs. There was one person left in the office. He seemed reasonable and said he would put it on my declaration, but I should notify Milwaukee customs upon arrival.

I did as he asked. Milwaukee customs confiscated the diamond and said they had the right since it was smuggled into the U.S. I was devastated.

After a proper welcome home with Jan and the family, I headed down to customs to talk to the head of the department. There was no indication they would relent, but I filled out voluminous forms detailing the search and rescue, the flat tire landing and the call back to Ft. Lauderdale with their agent's name. They took all the information and said they would get back to me in due course, whatever the hell that meant. For the first time in my life, I was sorry I volunteered to do a search and rescue. I was mad.

Three days later, I was summoned to the Custom House. I met the head of the department, the one I had talked to when I first arrived. As I entered he stood up and shook hands as he introduced himself. I was wary but kept quiet.

"We checked on every detail you gave us, from the San Juan request for search and rescue, your 30-minute gas stop in Nassau, your flat tire, and I personally talked to the agent, Harrison, whom you called from your plane. He confirmed that you two guys looked pretty beat up after flying low over the ocean for 3 ½ hours and then experienced a scary landing with a flat tire. It all checked out. We're giving you your diamond back. Next time have it shipped." He handed me a package.

I tried to explain why I did it the way I did. He was interested enough to hear the whole story and more about life in South America. He had read some of the articles in the Milwaukee Journal, so he wasn't completely out of the loop as to what I was doing. We shook hands. I thanked him again and was off to the diamond cutter.

Jan and I stayed up most of that night catching up on all the past news, the latest on the drowning deaths of Bill and Wes, what was happening with our four children, their schools and health. Then I dropped the bomb about having a house rented in Grenada and Diney, Tom, and Todd, enrolled in school down there for the coming semester. That took up the rest of the night. Life was not dull and not surprisingly, Jan sounded enthusiastic about our new home in the West Indies.

Donald Haack

Don grading 7 ½ ct rough diamond.

16. PRESSING ON

The unfortunate incident with Patlow hit us a double blow: a big chunk of cash flow disappeared and the loss of the second airplane for access into rivers not accessible by our land plane. Our priority was to place another seaplane in service, but that took more financing than we had. That meant we had to get more cash and the best way to do that was to fly the Cessna. The business was there. I had the plane and the flying ability.

I took every charter going into the interior and picked up return passengers by rotating stops at the airstrip trade shops, and leaving schedules of return days. It gave the miners a dependable timetable to plan their trips to town and almost doubled the flying income. I flew a grueling 100 hours a month and helped with the maintenance in the evening in order to have the plane airworthy for the early morning flights.

I thought we pretty well had all the bugs and problems worked out from the irresponsible lack of maintenance by Jimmy, but new ones continually popped up. The worst discovery was when I landed at the Madia airstrip where I unloaded three passengers and was heading to Georgetown for a second trip. That was not to be.

After dropping off the passengers, I taxied a hundred yards for take-off. Madia was one of the few DC-3 strips built a long time ago for the gold mining companies that operated in the area. Most of the easy gold had been mined and what was left was difficult and too expensive to extract. The mining company closed up but kept the big maintenance sheds intact, in case the price of gold rose to where it would once again be productive to mine.

I performed the usual run-up, checked the magnetos for any significant rpm drop. With everything in limits, I shoved the throttle in full for take-off rpm and accelerated down the runway. I lifted off at an easy 65 mph, pulled the throttle back into the black cruising and climb range, and watched the altimeter record a 500 fpm—*feet per minute*—climb rate. I sat back and relaxed as I trimmed the plane and set the course on 85 degrees to Atkinson Field.

Then it happened—a loud bang followed by a tremor and then a strange pulsation affecting the whole airplane. I immediately pulled back on the throttle. I didn't know if I flew into something or had structural damage. All I knew, there *was* a problem. I cut back power and speed, reducing stress on the engine and airframe by square root factors. I scanned the wings and prop and instruments for a clue. Nothing.

After the unknown noise, I couldn't believe what I was seeing—the engine fuselage and prop were moving up and down in slow motion as if not connected to the rest of the plane. I had turned back to the airstrip and with reduced power and gliding could easily reach it for a landing. What I was seeing was beyond belief—part of the plane, the prop and cowling around it were moving as if not attached to the main frame of the plane. I had the feeling I was watching a comic strip . . . only this was not funny. I fully expected to see the engine leave the plane altogether. Planes don't fly out of balance and I could visualize the plane falling out of the sky. I reduced power as much as I dared in order to keep from stalling and yet have a safe margin to land at the airstrip.

With the low power and airspeed, the engine movement lessened. When I was sure I could land in a glide, I cut the power altogether. Touching down, the engine took a perceptible dip down and then up. Something drastic had happened. I taxied back to the shed at the slowest possible creep to keep the engine from bouncing excessively. With the cowling off, the problem was evident: the left engine mount was broken and more than an inch out of line. A couple of minutes more of flying and the engine would have broken away from the frame…and I would have become a statistic. I gave a silent thanks to somebody for watching over me.

Jimmy must have abused the plane on overloads and hard landings and the airframe simply gave out. Now the problem was mine. I was a long way from civilization. I had a wild plan.

I found a straight stick and carved it until I could insert it into the upper section of the broken frame strut. It went up eight inches and into the bottom three. I figured if I could find a pipe, shove it up as far up as it could go, then line up the frame and let the insert pipe drop down into the bottom three-inch stub, I would have a firm guide for the frame.

I took the stick and went to the shed. The door was locked but windows were open so I crawled in. The building was huge—an old machine shop. No matching pipe, but I finally found a piece of solid steel the same dimension as my stick. I returned to the plane and inserted it into the frame strut. It was a good fit, but a foot too long. I found a steel hacksaw and plenty of blades in one of the drawers. There was a vise on the bench to hold the rod while I painstakingly cut through the ¾ inch steel rod. Fortunately it was iron and not tempered or stainless steel, which would have been impossible for me to cut. Even constantly using oil on the blade for lubricant, my arm felt like lead by the time I finished. I took my masterful creation, found a file, rounded out both ends of the steel rod and cleaned it thoroughly.

This is where I could have used a couple of more arms. I had to structure a *Rube Goldberg* design. I found a solid 2 x 2 inch board and maneuvered one end in the engine frame and the other on my shoulder. Using the 2 x2 for leverage, I guided and lifted the engine enough to insert the pipe into the upper half. I had marked the lower three inches of the pipe in chalk, so when I lined up the engine mount and the solid iron dropped into the lower frame, I could see the chalk and knew when it seated all the way down. That turned out to be the hardest part. It did not accommodate me by dropping down. It stuck part way. I took out my knife and by jiggling the frame with my shoulder on the 2 x 2's and using the knife blade in the ½ inch space between the broken strut, I gradually eased the bar down until I saw the chalk. Voila, a solid joint.

Well, not quite. That was the first step—to keep the frame from any lateral movement, but it could still move up and down with engine thrust. That's where my handy baling wire came into play— to hold my patch in place. The 400 pound tensile strength with eight wrappings meant I had 3,200 pounds of strength holding the frame from pulling apart. The inside solid steel bar eliminated any side movement.

I crawled back into the shed to return the tools and the left-over steel bar, and a twenty-dollar bill with a note as to what I had done and to call me in town if more payment was needed. Never heard from anyone.

I secured all the paneling on the plane, did a run-up to full power and the engine sat like the Rock of Gibraltar.

I flew into Ogle because I needed an American A & E mechanic to write off any major repairs. Harry Wendt took one look at it and smiled.

"You've really gotten into these bush repairs, haven't you? Not bad, really. Esthetically, it looks like hell, and we'll have to do something about the baling wire, but structurally, you weren't at risk—not bad." Coming from the old curmudgeon who was the best A & E I have ever known, that was quite an accolade.

He removed the baling wire, welded three small bars across the break, neatly trimmed it up, and wrote up the repair in my log book. I was back in the flying business.

Donald Haack

17. AN UNEXPECTED FRIENDSHIP

Mining towns, whether in the California gold rush, the Alaskan gold frontier, or the South American diamond areas, were the focal points for miners on their days off. They were places to relax after the long hard hours in the bush where the hope of big riches kept the risk of injury or death high. Days off provided a needed release of tensions and recuperation aided and abetted by hard drinking. Georgetown, British Guiana, was one of those places.

Aircraft maintenance and a much-needed buying trip for camp supplies brought Frank and me in for a couple of days. Frank could hardly wait to tell me, "We've been invited to Rodriguez' Portuguese Hotel for a fete they're having tonight. Lots of our flying customers will be there and the hotel is featuring East Indian cuisine. Drinks are free and the cover charge is ten dollars, a bargain, and we should go. Besides the regular miners, we might meet some interesting locals."

The fete was one of the bigger ones. There was a band going full time, people dancing and a line of tables loaded with curry, *rotis*, *dal puris*, and other East Indian dishes I hadn't encountered before. We sampled everything and washed it down with copious amounts of the local Banks Beer. East Indians, both men and women, introduced themselves to us with suggestions on what curry dishes to try, some of which were very hot, others mild. They showed us how we could alter the *horsepower* with combinations of the spices on the table. We met some very wonderful people and were introduced to out-of-this-world East Indian cooking.

Frank was particularly enjoying it. He found an East Indian lady who was fascinated by his stories, enough to put up with his limited dancing abilities. I noticed in the relaxed atmosphere of the party, his sometimes troublesome leg was not giving him problems tonight. The rows of chairs around the perimeter of the dance floor provided a constant exchange of partners. The band played without interruption. I sat there taking in the show, noticing that the young ladies were going full stop. The men escorted the ladies back to their seats after every dance, but the ladies never had a chance to recover before they were snatched up by the next admirer. I was amazed at their endurance.

I noticed one exception—an East Indian lady, fairly young, sitting alone. No one asked her to dance. After a while it became painfully obvious that she was not going to be asked. I couldn't figure it out. She wasn't a raving beauty, her dress was kind of full and frumpy, making her appear older than she was, but she had a regal air about her and handsome classic features that set her apart. It didn't seem to bother her that no one requested a dance. The thought crossed my mind that she might have some handicap, but I ruled that out when one of her friends sat alongside, and she quickly got up and moved to make room for the other's dancing partner who needed a much-needed break.

I'm the one at weddings who, when no one else asks, dances with the mother-in-law, elderly aunts, and widowed ladies whose foot tappings indicate their desire to hoof it on the dance floor. I walked over to the neglected lady, sat down beside her, and introduced myself as best I could above the clamor of the band. She extended her hand, smiled a big beautiful smile and said something that sounded like Shakira, which I took to be her name. It was too loud to talk so I indicated a dance, which she accepted. She was tall and when standing, looked anything but frumpy— and danced like a dream. It was a pleasant dance, after which I walked her back to her chair. Since no one came to claim her for the next dance, I sat down alongside again, even more perplexed why she was being ignored by the other young men. Later I would find out.

There were more dances with her and when the music was more subdued we were able to talk. I was surprised at her wit and sense of humor. We had an unexpectedly fun evening. It was late, the crowd thinned, and I said I would stick around until her ride or escort came to take her home. She scanned the room.

She touched my arm lightly. "My friends seemed to have been carried away with the *feting*, and I don't know when it will occur to them they left me behind. Would you mind terribly to take me home? Getting a taxi at this hour is almost impossible and walking the streets of Georgetown at night is out of the question. Would you?"

I drove her home and we talked for quite some time in front of her house before we decided it was time to go. We agreed it was a fun evening and hoped to meet again some time.

Two weeks later Frank and I were back in town. "Hey Buddy, the lady I met at that fete a fortnight ago has invited us to come to an East Indian party at her house. She said it was a group of friends, not a couples party, and mentioned that the girl you met, Shakira, would be there and hoped you would come. I would like to go, but I'd feel a lot better if you came along."

We went and it was entirely different than the last fete. There was no band, just soft music in the background, so conversation was possible. It was a welcoming group and we were able to meet and talk to everyone during the course of the evening. Shakira was there and made a point of introducing me to all her friends. Everyone knew her. We drifted apart and then back together several times throughout the evening. No one drank too much. It was an enjoyable evening in Georgetown with an interesting set of young people.

I thanked our host and was about to leave when Shakira came up to the door. I was about to express my thanks and say goodnight, but added, "You do have a way home tonight or can I offer the taxi service again?"

"Thank you, I'd be delighted."

Donald Haack

We left and discussed politics and politicians and the future of BG as we sat in the car in front of her house. I finally got around to the perplexing question of why none of the local young men danced with her at the last party. She laughed. "I guess you really don't know. I won the Miss Guiana contest two months ago and I do believe the title intimidates these young unsophisticated men, and they shy away from me." I shook my head and said I thought the logical reaction would be a long line of suitors standing outside her door. I didn't understand.

There were several more house parties that Frank and I were invited to and since we didn't have any place to reciprocate, we agreed to bring some of the food and drinks so we weren't free-loading. Even though it was a group and not a couples' party, Frank homed in on his lady friend. That had all the aspects of a *serious* friendship. More by default than design, Shakira and I spent a fair amount of time together in discussions of the growing ominous trend of the division between East Indians and blacks. The country was always known as *one country, one people*, but with independence coming up, the politicians were stirring the pot on racism—the *divide and conquer* theory. Only it was taking on an ugly dimension never experienced before. The phrase *choke and rob* became part of the daily vocabulary. The unemployed blacks were brazenly attacking East Indians, businessmen, tourists, or foreigners who wandered the streets at night and, lately, more frequently during the day.

My existing friends were mostly business owners: Eric Schrier, from Switzerland who owned the Swiss House jewelry store; Mac Wilshire and Alan Humphrey, owners of the two prominent hotels, the Woodbine and the Tower Hotel; Dennis Khouri who's family owned the Khouri Department Store; Les Ricketts, a partner in Wilson & Ricketts, the biggest construction company in the country; and others who mainly felt that the growing schism caused by the politicians might explode into wrecking the business community. Shakira and her friends were concerned about events far exceeding the business aspects—the security of the East Indians.

Most of the East Indians were from middle class families and their main concern was being targeted and victimized more and more. The government, unable to stop the crime wave, was the main fear of the East Indian community. Ninety-five percent of the police force consisted of blacks. The few East Indian police were relegated to minor positions. Promotions for them were far and few between. Shakira and her friends provided interesting but ominous discussions on the future of BG.

On my next trip to town there was a note to call Shakira. "My friends have recommended a good movie that's playing all week— 100% East Indian production and actors, and think it would be a good addition to your cultural education to see it." I heard about those movies. They didn't sound like my taste, but I reluctantly agreed to go. "Can you stop by to pick me up at 6:30 for the early show, and we can eat later?"

Shakira was waiting outside when I arrived.

"Are we meeting the rest of our gang at the movie house?" I asked.

"No, they've all seen it already, so it's just us." My reaction was a mixed one, but decidedly uncomfortable. When I first met this young lady, she appeared a bit frumpy and socially left out of the circuit, and I had no problem with my conscience. But after a few gatherings, it was evident that she was socially adept, sophisticated, intelligent, and handsome to the point of being beautiful. In all our meetings we were as a group, not as a couple. The two of us going to the movies took on the unexpected aspect of a date.

"You're quiet tonight, is everything all right? Oh, we're here. Pull in front of that car where it's light; less chance of someone trying to break in."

I pulled in and whatever I was going to say would have to wait until later. I bought the tickets and we went upstairs in the balcony. No one of means sits downstairs in what is termed *the pit.* Those seats, or more correctly benches, sold for ten cents and were noisy with occasional fights. All movie houses in BG had that make up.

East Indian movies are highly emotional and much over-dramatized. During one of the more emotional peaks, Shakira clutched my hand in hers. My adrenalin shot up but in glancing over it didn't appear to affect her the same. She was very intent on the sword-swinging hero saving a damsel. After the hero's crises had passed, we were still holding hands. My reaction fortified my resolve to talk to her later.

I squeezed her hand to get her attention. "Shakira, there's something I have to talk to you about...after the movie."

"Okay, but my, you look so serious. This film isn't getting to you is it?" she added playfully.

"No, no, not that," I mumbled.

After the highly emotionally charged movie, we drove down the street for a Chinese dinner. I stopped her before she got out. "There is something I have to tell you. It wasn't important before and there was no occasion to bring it up. But tonight is the first time we've gone out together—alone. Like a date. Shakira, I'm married, in fact very married with four children, and happily married and—" she put her finger on my lips to stop my rambling. I felt like a little boy.

"Silly, I know that. I found out from John Partin after the first night you took me home. John explained your situation, particularly that you didn't get involved with any local women and that your wife and family were in Grenada. And, thanks for telling me about your status. It makes me feel good. I guess if you hadn't it would have changed the relationship between us. Everything's fine. We have a close friendship and we can keep that. The only kisses I get are the very same social pecks on the cheek that everyone gets down here. And as far as holding hands, I like it. And you've been down here long enough to know it is an intimate but very proper custom that even the men do together. You told me how uncertain you felt the first time your attorney, Eric Clarke, held your hand as you walked down Main Street. Later, at the Ministry of Mines' meeting, the ministers, all men, and you were standing in a circle—holding hands during the discussion."

I was at a loss for words. "You can close your mouth now before we go in for dinner," she laughed. "You know I leave for London in a couple of weeks so we don't have much more time together."

A few days later I called her. "Shakira, I have a friend coming down from the States. He's been commissioned to do a thirty-minute film on my safari and the interior. Eddy has a last name, but I only know him as *Fat Eddy*, my brother's nickname. He's shooting this for Schlitz Brewing Company in Milwaukee and the Johnson & Johnson Wax company in Racine, Wisconsin. He asked if it would be possible to find a Miss Guiana, if they had such a title, and to use her for the opening shots, possibly in the Botanical Gardens. It just so happens, I know of a Miss Guiana. Would you like to be a part of it?"

She agreed and would bring along a friend to make it look more natural. The filming took the better part of one day before the filming crew and I left for the interior.

"Don't forget that date I gave you. It's important that you be here to escort my mother and me to the Miss Guiana Government reception. It's a kind of send-off fete for my London trip." I said I would be there.

When I returned for the government reception, I had some misgivings about escorting her to the function. "No, it's all very formal and someone has to present me to the Prime Minister. It's not me, only. You'll be escorting mother, too. You can be *her date* if that would make you feel better." I rolled my eyes. It was hard resisting her plans once they were set.

That evening everything was going according to plan until it was time to leave the house. Mom wasn't ready. "We're going to have to go. They have a reception line set up, and we can't keep them waiting. You'll have to come back for mother."

I didn't realize the full impact of that until we arrived at the building. There was a long reception line of all the ministers and everyone who was anyone in Georgetown. I knew most of them. Gawd-almighty! There I was escorting her, Miss Guiana, down the reception line as I tried to make small talk and quell raised eyebrows silently asking what in the world was I doing escorting her to this formal function? I couldn't address it, because I myself didn't have a clue. I don't blush easily, but I am sure my face was a dark shade of crimson. The line lasted forever. I finally pulled away.

"I'm going for my *date*, Shakira, to bring your Mom up before the reception line is down. I raced off. Fifteen minutes later Mom and I returned to an empty reception area. Everyone was inside. The damage was done.

Inside, there wasn't much of a choice. The tables were full and my name plate was at the head table, between Shakira and Mom, along with eight other dignitaries. I felt like the black swan. Some of my more crude friends sidled up to me with that wicked sneer to ask how I managed to do that, whatever *that* was. There was absolutely nothing I could say that wouldn't sound phony, so I said nothing and had another drink. After several of those, I must have mellowed enough to be sociable and invited Miss Guiana to several dances. *In for a penny, in for a pound* I kept thinking as I danced with Miss Divine. I couldn't help reflect that dancing with her in the first place is what got me into this situation. If I were single, this would be a heavenly experience and would have a different ending. Life can be confusing.

Towards the end of the evening, I drove Shakira home and did a quick turnaround to retrieve her mother. When I returned to the table I was met with indifference bordering on open hostility and was summarily informed *Mom* would find her own way home.

When I talked with Miss G a couple of days later, I remarked about the war-like attitude I received from her mom that night. "Oh that. I guess I neglected to inform Mother about your married status for the same reason you didn't tell me at first. There *was* no reason to. We were not dating. But you know how mothers can be. You and she got along fabulously well, and I think she had some deep and dark thoughts about you as a possible son-in-law." She shrugged her shoulders as if to say, "Don't ask me."

"The night of the government reception someone told her you were married and it hit her full force. We had a big argument about what was I doing running around with a married man. I tried to explain, but it fell on deaf ears. We're kind of in a neutral standoff at the moment, but it, too, will pass."

I made a point of making myself scarce. Her mother thought I was a terrible ogre. Two days before her London trip, Shakira called. "Can you pick me up tonight at six? My close friend, Oscar Greene, the minister, invited me to his home for a going-away-to-London send-off. He said to bring you, too."

The house was packed with well-wishers; the food and drinks flowed along with the warm camaraderie. I had a close feeling with these people and would miss the interaction— as I would with Shakira. There were no places to sit down. Shakira steered me to an empty floor space. We sat down on the floor, ate, and held hands. We knew it would be the last time we would see each other. When I took her back to her house, it was a short stop so as not to encounter the wrath of Mom. We had our first kiss, a good-by one. No dry eyes. I drove off thinking it was like two ships passing in the night. There was nothing to be ashamed of, but it would be an indelibly carved good feeling in the passage of life.

Later, I tried to reflect on our situation, but it wasn't easy. Our original meeting was by happenstance, not by design, and a good friendship evolved. She was headed to London and the glittering Miss World contest. I would be spending some of the time back in the bars listening to drunken miners. It was a depressing thought. Shakira had outgrown Guyana, and I was sure she would stay in the UK. I hoped whoever married her would appreciate what he had and would treat her accordingly.

A couple of months later it was announced: Shakira was chosen number two in the Miss World contest. She was on her way. Two years went by before I heard anything more—an article from my brother in the States. Shakira had played a couple of roles in movies, one with Michael Caine, whom she married. Eventually they had four children, and it was one of the rare movieland marriages that lasted a lifetime without a divorce. Good for you Michael!

Donald Haack

18. FLASHBACKS

To get our income back on an even keel, my flying was extensive and included trips back to the Rupununi where we lived when we first came to South America. On the way to Good Hope from Kamarang, I decided to deviate off course and pass over our old home site. I shouldn't have. As I came over Marquis, the airstrip was easily visible. The house we built where Jan, Diana, Tom, and I spent our first five years together was a shambles. The roof was half torn off, one of the walls had disintegrated, and the water tower that supplied our house with water was no longer standing. Dutch's baracoun that housed our staff and the company store were barely distinguishable. Nature was taking Marquis back to what it was before we moved in nine years ago. The hangar had fallen down but, surprisingly, the frame of the Super Cub airplane we salvaged from our crashed landing was still there.

It was an intensely depressing sight, particularly because of all the emotions, crises, and tribulations Jan and I shared in that home. I could almost picture Diana feeding our pet deer, Bambi, and Dutch walking up from the store to our house for his evening drink. The memories and tears were blocking my vision.

My last flights out of Marquis to Tipuru airstrip alongside the Tipuru River, came back vividly. The airstrip, a couple of hundred feet long, adjacent to the river and ending at the base of a thirty-foot high stone outcropping, was sub-marginal—much too short. In order to land in that small distance, I drained all but a few gallons of gas from the plane, took out the seats, carried a half load, and only attempted landings at the break of dawn, before any hint of a breeze. A tail wind would increase ground speed and the distance it would take to stop. Every flight was rehearsed a hundred times in my mind the nights before and each successful flight added to the stress to the point that I hardly slept before attempting the trip. With zero margin for error, I tried not to think of the odds I was playing against.

On the eight minute trip to Tipuru, I flew a few feet above the river at slightly above stall speed and when the airstrip was in sight, I applied full flaps and full power. If everything went as planned, the plane wouldn't stall because it was kept buoyant by the lift from the propeller wash. At the last second I cut the power completely. The plane had enough residual lift to rise to the level of the strip where it stalled in the first few feet. I stood on the brakes until the plane stopped within spitting distance of the rock escarpment and I collapsed mentally and physically. I knew these landings were outside the plane's design limits.

One morning after a sleepless night before my planned flight to Tipuru, I met with more unexpected stress. Daylight was breaking and there was no breeze—a requirement for this flight. As I walked along the path to the airstrip, I noticed a slight movement to my left—a coiled rattler raising his head. Instinctively I didn't step down, but skipped a bit forward with the foot that was the closest. The rattler struck at my leg, catching his teeth in the back of my khaki cuff. I kicked loose, slid out my pistol, and shot at his head, the only part of the snake that doesn't move from side-to-side. The first round of birdshot I used for trick-shooting proved its merit. The snake lost its head. I was unnerved and considered quitting for the day but I calmed down and changed my mind. *As long as I'm up, I might as well transport the badly needed pump motor to the camp.*

I checked the airplane thoroughly. Dutch had removed the seats the night before and drained gas to the minimum needed. Everything A-okay. I opened the plane's cargo door. I bent over to pick up the 5 HP water pump engine sitting alongside on the ground. My hand grabbed into the space between the three-inch legs. As I lifted I felt something soft between the steel and my fingers…and immediately pulled away. I stepped back, and the silent air was filled with an angry rattle. I must have squeezed the hell out of the rattler's poor belly when I lifted up. No wonder he was angry. Heck of a way to wake up. I could hardly believe it. In all the years in the bush, other than the bushmaster incident when my plane crashed, I never encountered snake problems. And now, two rattlesnakes in one morning. Maybe it was a sign.

I didn't shoot him. He was more disturbed than I was.

I hated this flight to Tipuru, what with lost sleep and a lifetime of adrenalin used up every time I landed there. I removed the keys and walked back to the house. That aborted flight was my last attempt to land at Tipuru airstrip. It was permanently closed as far as I was concerned.

Flying over parts of this territory was like opening old chapters—memories poured out. As I flew over the Karasabai airstrip and crossed the Ireng, I could see the outline of a strip I made eight years ago. It paralleled a footpath alongside a small cluster of Indian huts. I landed there several times with Doctor Talbot, the medical doctor from Georgetown. He made the rounds of villages twice a year and was the best medical doctor I worked with. He told me he had several careers before he decided to become a doctor. He enrolled in the London School of Medicine at age 40. He was by far the oldest in his class and said it took persistent arguments to win over the school authorities allowing him to enroll instead of the many young pre-med students. It was a wise choice.

Before Doc Talbot came along, I flew several younger doctors on the same route. Because my plane was small and the medical supplies and personal belongings of the doctors were extensive, there wasn't room for an assistant, nurse, or anesthetist. I became all three and that meant I had to administer ether while the doctor performed surgery, set arms, or pulled teeth. The first thing these doctors did was inform the *Tuchow*, chief, and the *Peayman*, medicine man, that they were not needed. The young trained doctors preferred to work alone, with the exception of me, the voluntary assistant.

I became adept at anesthetizing patients with ether—and not losing them. The trick was to release a few drops at a time on the gauze mask, then continue to monitor pulse and blood pressure. I had to convince the patient to breathe deeply and all would be well. But I had a guilt complex. I hated ether. It made me sick when I was a kid, and I knew the patient would be wretchedly sick afterwards. But ether was safe and easy to apply, the reason it was used extensively in bush procedures. If I could administer it, anyone could.

Then along comes Doc Talbot with an entirely different approach. The first time I flew him to a village he asked for the *Tuchow*, who happened to be gone for the day. He then requested the *Peayman*, and through a translator explained that he wanted the Peayman to work with us. Doc asked the Peayman to do his *hands on* procedure to calm the patient, and he in turn would show him his medicine. They could learn together. At first the Peaymen were skeptical, but with time they became great supporters of Doc Talbot. His rapport with the tribes was unlike any other doctor's. For me it was a real eye-opener to see how potent the Peayman was and to realize we could learn as much from him as he from us with our so-called modern medicine. It was an interesting blend of two cultures.

Working with the medicine man all I had to do was adjust the gauze mask on the patient. The Peayman stood on the other side of the patient, moving his hands a few inches over the body without touching. Sometimes he hummed. After a few passes with his hands, the patient relaxed and went to sleep. I was ready to apply ether if our patient didn't respond to the Peayman's treatment. I glanced over to check with Doc Talbot for guidance.

"I believe the Peayman has our friend in a deep enough sleep. We'll proceed with caution. If there are any signs of discomfort, use the ether, but I think we're okay." He continued to set the

Donald Haack

broken arm without a whimper from our patient. And so it went with them all. There was an exception, but that was because one Peayman was young, inexperienced, and still in the learning stage. He didn't have the confidence in himself or his fellow villagers, and I had to start up my ether drop. There was a moment of possible alarm, but it passed quickly and the procedure continued without a hitch.

Years later Jan and I were at a *Mashramani,* a meeting of the Wapashani clan in the United States. We were "adopted" into the tribe when we first lived in South America, and we continued to attend the Mashramanis whenever we could. This particular time, we were sitting in a big room, and there was a nervous little dog running around, yipping, barking and being a damnable nuisance.

Tandy, the young lady sitting beside me, was fun to talk to. I knew her mother from back in the Rupununi. It was hard to realize the last time I saw her she was a small bundle in my arms, and now a grown, beautiful young lady. The yippie dog came close, and Tandy picked it up and put it in her lap, holding it with one hand and stroking with the other. In seconds, the dog was quiet, then limp. She was stroking it with both hands now, and the dog was totally out, sound asleep. She picked it up and gently laid the sleeping dog on the carpet.

"Tandy, what did you do? I thought at first you were petting the dog, but you weren't touching him with your left hand…and he went out like a light. What happened?"

She smiled and shrugged as if it were nothing. I wasn't getting an answer.

"C'mon, Tandy, tell me what you were doing. You know I knew your mother, Edwina, and she had some very unusual healing powers. Did she pass some of that on to you?" I wasn't letting her off the hook, I really wanted to know.

It was obvious she was not comfortable talking about it, but when I told her about knowing her mother's abilities, she loosened up and told me the whole story.

"Yes, mom taught me what you just saw and quite a bit more. Most people stare in disbelief when I discuss it, but you lived down there and knew my mom, so you understand. I studied nursing when I came to the States, and after graduation got a job with the University of Michigan hospital. Some of my older patients were high-strung, stressed out, or had inefficient immune systems. I used my *hands on* approach to calm them down and lower their blood pressure. A doctor watched me do this several times and asked to hear the whole story. I was reluctant at first, but he seemed to genuinely want to know more about these Amerindian procedures.

Later he came to me with one of his patients, an elderly lady weak and in failing health. She needed surgery but they were afraid she wouldn't survive a general anesthetic. He asked me if I would assist so that she would use only a minimal anesthetic to put her under. I agreed. I did the *hands on.* She was responding well, but I guess they weren't all convinced. They still gave her a bit of anesthetic. It was almost too much, and they had to back off completely, while I kept her under. It was a successful beginning. After that, many of the doctors came to me with their marginal patients. They all turned out okay. That's when they decided to send me to Denver for a special training to be certified as a hands-on anesthetist. I'm being used full time in that capacity now."

"Well, I knew the University of Michigan was one of the *avant garde* institutions of learning, but it has just gone up several more notches in my estimation. Congratulations, Tandy. Your mom should be proud of you."

As I flew further into the Rupununi, the landscape below kept changing like the kaleidoscope of memories it released. Good Hope Ranch was coming up on my right. Caesar Gorinsky, one of my closest friends, still resides there, but his wife Nellie, Jan's closest friend, moved away. Caesar, with his big heart, had two basic faults: women and gambling. Nellie put up with both for twenty-five years, until both habits became intolerable. His Georgetown gambling forays mortgaged Good Hope to the point where it jeopardized the lease payments on their unique ninety nine-year land lease the government granted to ranchers. And then his womanizing got out of hand. As tolerant as Nellie was, she finally reached her breaking point and moved out. I intended to visit Caesar on my return trip but Good Hope would not be the same without Nellie. I could picture Nellie and Jan working and laughing together in the kitchen, baking bread or having tea in the living room with pigeons flying in to join them.

Eight years previously, while our house in Marquis was being built, we were guests living with the Gorinskys. Dinners were always a challenge—there are just so many ways to prepare beef and jerked meat. Nellie thought it would be a good time for Caesar and me to go on a duck hunt and bring back a change of diet. Jan asked if she could join us.

We were up at 4:30 A.M. A twenty-minute jeep ride took us to our first site: several big ponds with cashew trees along the edge. At dawn, Muscovy ducks, the size of geese, came in to land over the cashew trees where we were positioned. Six ducks was our self-imposed limit. After we had our limit, we moved on to the bigger ponds for the smaller ducks.

This one day Jan joined us but wanted no part in the shooting, just to be out and join the hunt. Seven months of pregnancy produced *cabin fever* and a need for change. Hopefully, our duck hunt would provide that.

Caesar pulled up the jeep alongside the big pond. "I'll cross over to the right side with one of the boys. You and Yan can go to the left. It isn't as far. She's *heavy with colt*, so it'll be easier for her." That was Caesar's expression for being pregnant, a phrase Nellie and Jan detested and gave him the *bad eye* when he used it.

"Now you'll see why we wear tennis shoes and long pants," I said to Jan as I extracted my Browning automatic 12-gauge shotgun out of the jeep and pointed to the water alongside. "We have to wade through the water to reach that point of land over there." It didn't seem far, but considering we would have to walk through waist-high water, it was enough of a walk. "Caesar, you can take both of the boys. Jan and I will be okay by ourselves."

"You're not going to make Yan fetch the ducks, I hope. You know she's heav—"

"No, no," I interrupted him. "I'll do the fetching and the shooting. No problem." And off we went. Jan stepped gingerly into the water, but then waded in right behind me, so she wouldn't be left by herself. In a few minutes we climbed up the bank onto the small island. I could barely make out Caesar across the water. He was a long way off, but as planned, when the ducks came in, he shot first. Disturbed on that side of the pond, the ducks circled to our side where we were waiting for them. Today's limit would be twenty ducks. There were no guests at the ranch now. That meant a welcomed menu change for a week of roast duck. If guests showed up, we'd have a treat for them—and then another hunt.

We were barely settled in behind a couple of three-foot high shrubs when we heard Caesar's first shot. Actually four shots. We didn't move and searched the horizon. Ducks were advancing fast along the left side of the pond. I waited. The first ones came in five feet off the water at sixty miles an hour. Several passed before I was ready. Another four were coming up. At that speed I figured to lead them by four feet. I shot twice. Nothing. Another six ducks. This time, leading by six feet, I shot twice—one duck plunged into the water less than twenty feet away. A six-foot lead was the magic number. Two more ducks approached—one shot, one duck down. Now they came in fast, faster than I recalled in past hunts. I wasn't doing too well—five shots, only two ducks. Caesar would have a derogatory comment about that. As in most of his endeavors: playing chess, hunting, and gambling, he didn't like losing. He abhorred losing. God help me if I shot more ducks than he did or had less shots per duck than he. That wouldn't be the case today. I was off to a slow start.

There was a pause for a moment. No ducks on the horizon. No shooting on Ceasar's side. I took that opportunity to hand the shotgun to Jan. "Hold this for a minute. I have to fetch the ducks and set them up as decoys." I ran to the water's edge and retrieved the two ducks. I placed them in a sitting position and held up their heads with a couple of small sticks. It was as if they just landed. I

Donald Haack

ran back to our blind and Jan handed me the gun. In a few minutes there were a couple of more shots from across the lake and another flock of ducks coming at us. This time I was ready—two shots at the leading two. They both dropped. Two-for-two, not bad. Then another one-for-one. They were still coming in. I handed the 12-gauge to Jan. "Careful, the barrel is hot. Be right back."

"Why don't you wait? There are more ducks coming in." I was out and back in less than a minute while ducks were passing overhead. She repeated her question. "Why don't you just wait until you have several more of them in the water and then retrieve them all at once? Would make a lot more sense than running out in the water for one or two ducks each time." She turned to me quizzically for the answer.

I hesitated. "Well, you can't leave ducks in the water very long. The alligators will get them. If they get into a feeding frenzy, it's best not to be in the water and you lose all the ducks."

Her quizzical expression turned to one of incredulous discovery. "You mean there are alligators in there…in that same water we walked through to get here?" Her voice got perceptibly higher and strained. I continued to shoot and retrieve until I counted ten ducks. We had decided on twenty, but I would bet the farm Caesar would have more. He had to do the one-upmanship or he'd be in a foul mood the rest of the day. As long as we all understood that, it would be a pleasant outing. First-time hunters with Caesar who didn't play by the unwritten rules would by the end of the day wishing they had.

In spite of my trying to dispel the alligator fears to Jan, she constantly looked around. "Maybe I should go now while you're still shooting and take the long way back without going in the water. I'd much rather do that than return the way we came in." It came across as a hopeful question.

"Not much chance. We're on an island. Only way back is the way we came." It wasn't hard to tell that was not what she wanted to hear. "We can leave right now. We have our share of ducks." A long silence followed. I pulled a flour sack from my pocket, stuffed the ducks in and slung it over my shoulder. "Let's go. I don't hear any more shooting, so Caesar must be finished."

We waded into the water, only this time I not only had the sack pressing on my shoulder, I could feel Jan's pregnant belly pressed hard into the small of my back, so close that we were walking lock step through the water. You couldn't have pressed a quarter between us. We were so tight anyone seeing us would have thought we were one fat man crossing the pond. We crossed in silence, which was quickly broken when we reached dry land. In no uncertain terms Jan gave me her thoughts on my "unthinking calloused" indifference to her feelings and condition. I listened contritely and in silence, while I was thinking I should have known better than to bring a pregnant woman on a rugged duck hunt in the Rupununi. Men learn so slowly.

As I flew, flashbacks kept coming up like volcanic eruptions. Lethem was my destination. My flight path was far east of Good Hope where I could actually see the Karanambo airstrip near McTurk's ranch, Dutch's second Rupununi home. He planned his trips to arrive at McTurk's before tea-time to enjoy Connie McTurk's and Anton's (their cook), tea and cakes. They were out of this world. The setting was unique. Part of the house's walls had fallen down and remained rubble. A rough-planked dining room table was the focal point of the room, one side of which had no wall. For meals and tea the table was covered with a fine English linen tablecloth. Instead of chairs, orange crates provided the seats. In a most incongruous fashion, beautiful silverware and china dishes were at place settings. Connie's teas were to die for. It was no wonder Dutch always arrived in time for those memorable events. He would stay for a couple of months, then pick up his walking stick and cross the savannahs on foot. It took him two days and thirty miles to reach his overnight stop, Good Hope outstation where the foreman and family lived. They fed Dutch, after which he would hang his hammock, climb in, sleep soundly. In the morning he would be gone.

Donald Haack

The first time the McTurks invited us for an overnight, I loaded the plane with avocados I brought back from Tezerek's airstrip, which he called Velgrad. His avocados were a delicacy and appreciated; one of the few gifts I could bring the self-sufficient McTurks.

After Dutch flew his first trip with me to McTurks, he was determined to build an airstrip alongside the ranch house instead of landing on the DC-3 strip, five miles and one swamp away from the ranch. Not too long thereafter, he completed it, and we used it frequently, particularly at tea time. In the harsh and primitive bush life, small amenities became necessary *divertissements*.

The first time we overnighted, the McTurks insisted we sleep in the big double bed with a mosquito net surrounding it. The night brought bats, some of which carried rabies. Several cases were reported to the Lethem medical station every year. We obediently tucked in the mosquito netting, heeding the warning not to let our toes touch the netting because vampire bats could suck blood through the netting. If that happened it required a series of rabies shots—needles stuck into the stomach, a truly unpleasant experience. We kept our feet far from the netting.

I learned most of my bush mechanic repairs from "Tiny" who maintained his own engines and equipment. I followed him around while he listened to my frustrations about not receiving expandable rings for my compressors and small engines. The compressors were used seven days a week. Some of the Briggs & Stratton 3 hp engines had over a thousand hours with little compression left in the cylinders. He laughed. "I thought you had serious problems. Come, I'll show you how to get them back to new in no time."

We went to his shed where he was working on no less than a dozen small engines in different states of repair. He picked up a cylinder. "Look, these rings are worn way down and wouldn't give enough compression to even start the engine, much less keep it running. I expand the ring with this little tool I built. You can have it. I have several. Remove the ring. These English biscuit tins are my salvation. Connie saves them for me. See the heavy sealing foil on top the tin? With my tin snips I cut narrow strips that fit all the way around and under the rings. Then I replace the rings, and bingo— instant expansion rings and plenty compression. They may not go for a thousand hours, but neither will original rings. So there you have it…instant repair and as good as new.

"Oil and gasoline lines that leaked gave me the biggest trouble. I couldn't get replacement lines or the proper attachment fittings. I remembered an old WWI trick we used on our Sopwith Camel planes. Some feller in our squadron figured out a way to repair pin holes or cracks in oil and petrol lines. He used bar soap and thread. Rubbed a bit of bar soap on the area, then wrapped it tightly with thread, put a few drops of gasoline on it, which formed a seal, then repeated the procedure until there was a quarter-inch build up. An hour later it was functional and good as new. Look at my jeep." I did. There were lumps around every line. Looked like an old *camoodie* snake after it finished its meal.

"When we had problems with magnetos, generators, or starters shorting out, we packed them in an oven at medium heat for a couple of hours. The cooks hated it, but when we told them the place it played in winning the war, they gave us the use of the ovens anytime we requested."

Thanks to McTurk, I used these solutions many times, both on the airplanes and the small prospecting and dredging engines. When I tell this to an engineer, I get a strange, disbelieving stare as if to say, "You really aren't serious, are you?" Not accepted procedures by the book. But then neither is flying and operating equipment hundreds of miles from civilization where there are no certified and fully-equipped repair shops.

Donald Haack

Don and the Tripacer

19. WHAT'S FOR DINNER?

We had one particular safari of gung-ho guys willing to try anything new or adventurous. When they heard we were eventually going to move up the Mazzaruni River, north of Kamarang, to one of the hottest diamond areas, they offered to make the move themselves.

We hadn't done that before. We always had the camp settled before the guests arrived, but these guys were so enthusiastic about the idea that we went along with it. We sent Frank up river first with one of the big dugouts so that he would have the cook shack up and ready when the guests arrived with the dredge. Traveling isn't so bad if the food is good so that was number one priority. The eight guests would accompany the barge with the suction dredge on it. With them was our crew of four, led by a Frenchman, Jean Pierre. I felt the group was in good hands and went with Frank to make sure the camp site would be ready when they all arrived the next evening. We had two Arawak Indians with us, and within three hours they cleared an area for the guests' sleeping tents, while Frank and I set up the cook shack and dining tent.

We worked non-stop until two o'clock. Exhausted, we had lunch and took our much-needed siesta. Frank tied one end of his hammock on the same small tree as my hammock. I felt a bunch of jerking motions twenty minutes after we settled in. They were followed by Frank proclaiming he was *going after our dinner*, whatever the hell that meant. I was still half asleep, so it didn't register what he had in mind. At that point it didn't make any difference—until a few moments later when all hell broke loose.

It sounded like a war—there were a half-dozen rifle shots. That got my attention. In a matter of seconds, I was out of the hammock with my pistol strapped on and headed toward the gunfire. I didn't know what to expect so I proceeded with caution. I couldn't imagine what Frank would be shooting at. The Arawak Indians were friendly and if Frank wasn't firing in self defense, why was he was firing so many shots? I came to a clearing. Frank was running back and forth, his rifle pointed up in the trees. Every few seconds he fired off a round then scurried to a new position.

"Frank, what's going on?"

"Big monkey up there. Fresh camp meat for tonight," he gasped as he jumped over logs and dodged trees. When Frank was excited his bad leg worsened into a serious limp. That was happening now. With his bad leg, his pure white hair, and leaping around the forest floor, Frank looked like a *Troll* from children's books. I watched these antics for several minutes. For whatever reason, the monkey wasn't leaving the area. Every time Frank shot, the monkey jumped to another tree, somehow avoiding being hit.

"Gotta get camp meat for tonight," he repeated as he jumped over stumps, stopping only long enough to get a shot off. It was like a cartoon scene. Frank stumbled and fell several times. I worried he would injure himself and this wasn't the place for a medical emergency.

The monkey landed on a main branch where I had a good view. I slipped out my pistol, aimed a fraction behind the ear, and pulled the trigger. Seconds later the howler monkey came tumbling down. Frank was still running around in a circle.

"Frank, he's hit and coming down—right above you!" Frank saw him but not in time to get completely out of the way. The monkey crashed through a small tree, luckily breaking his fall before landing a glancing blow on Frank's back. Frank and the howler monkey tumbled to the ground in a big heap. "Are you all right?" I asked, running over.

Donald Haack

Frank was gasping for breath. The wind was knocked out of him and it took several seconds before he could answer. "I'm okay," he finally got out. "Just getting my breath—knocked the wind out of me. Did we get it? How did you get him with your pistol and I couldn't with my rifle?"

Domingo, standing alongside was the Amerindian from the Rupununi who taught me how to shoot by pointing. He looked at me and smiled. I was his good student. Frank wouldn't understand. Then I noticed movement from the monkey. Domingo walked over and held a tiny howler baby monkey still hanging on to its mother's back. I felt terrible. Frank was elated—He would cook the howler monkey for dinner and make a pet of its young one. I lost my appetite.

By the time the crew and guests arrived by barge, Frank had the big pot boiling and home-made bread to dip in the pot or eat with meat and gravy. It smelled absolutely delicious. Everyone was ravenously hungry and it seemed to be *a feast for a king*. That is until David, our fastidious guest from Chicago, scooped lower into the pot for more meat and a human-looking hand appeared in the ladle. Everyone stopped eating and stared. David dropped it back into the pot, turned around, rushed a few feet away and threw up. It was a mini catastrophe. Copious amounts of dessert rum helped the situation.

Frank was devastated that his dinner wasn't considered a success just because he left the monkey's hand in the stew. Eventually, he got over it and our guests *kind of* got over it. That campsite had its share of strange happenings. Frank took on the mother role of raising the baby howler monkey. The Indians informed Frank that he couldn't housebreak a monkey and of course that was a challenge for him. He had his own way of teaching. Every time the little monkey pooped in the camp area, Frank picked him up and threw him in the river. Monkeys swim but not well. When the little chap struggled back to shore, Frank would pick him up and dried him off. It only took a couple of days of being tossed in the water that our little howler figured out it was better to do his business outside of the tent area and away from Frank, than being airborne and doing a splashdown in the Mazaruni River.

Having gotten over that hurdle, Frank allowed the howler the run of the camp. It was quite a sight seeing him in the kitchen watching Frank's every move and sometimes tasting the food, if Frank wasn't looking. He even sat on Frank's shoulder at times, which made the picture complete—the wild-man-troll effect with monkey on shoulder. Frank envisioned himself as the original bush man, taming animals as they did on African safaris. This love affair lasted two months. The howler had his own little house where he slept every night. Frank inadvertently placed a fiberglass hardening chemical on the shelf above the howler's house. The plastic bottle had a crack, the chemical leaked over the house, and the fumes did in our pet. Frank grieved for weeks.

During the same time Frank came to me and asked if it would be okay to hire one of the Arawak Indian girls to help out with the cooking, the clean-up of dishes, and the dining area. She convinced him that she had experience and would be able to start right away. The camp she had worked in had broken up, the miners had left for town, and she wanted a job and money. I agreed. That turned out to be a good choice. She was a good worker.

But a few days later, we had another problem. Some of the local Indians took a liking to our camp cook and tried to enter her tent at night. Frank, the knight in shining armor, came to the rescue and asked if it would be all right if we allowed Marie to sleep on a cot in our tent. I thought that was a bit much, but Frank convinced me it wouldn't affect our privacy. She would only arrive after we turned in for the evening and would be out before we got up. She was barely noticeable and I forgot she was a night guest.

A few nights later, we had an incident. It was a full moon and we turned out the Tillie lamp to go to sleep. I awoke later with the feeling we were being stalked, and it took only seconds to confirm my fear. The bright moon illuminated the area surrounding the tent. Not two feet away from my cot, a figure moved alongside the tent, as if trying to find the entrance. He went back and forth along my side, totally unaware of how visible the moon made him. I slipped my pistol from under the cot and silently put a round in the chamber. Whatever he was doing out there was not the smart thing to do in the middle of the bush. I reached over and touched Frank and put my hand on his lips. He was wide-awake in seconds. He leaned over and whispered, "How long has this guy been watching us? What do you want to do? He's holding a knife. Should we take him out, or what?"

"Let's see what he really wants. We have the advantage of seeing his every movement, and he can't see us. If he tries to come in, I may fire a warning shot. That sure as hell will scare him off. If it doesn't, we take him out before he gets in. He can't see us, and we've got him in our sights already. Let's wait and see what happens." It was a bit unnerving watching the shadowy figure move around the tent. It came back to my side and couldn't have been more than two feet from my cot, as he put his hands on the tent, looking for a way in. I had my doubts for a minute whether he could see us, it was so close, but I assured myself that would be impossible. I kept the pistol trained on him as he crept around the tent. Frank kept looking over to me for a cue on what to do. He was more than a bit nervous about this intruder. There was no question about a knife in his hand. It was easy to see when he came close to the tent, which was almost all the time now. I slipped off the safety and was ready to fire.

He came to the front of the tent. His hands went up and down as if trying to find an opening. He bent down and the knife came up to the tent. I tapped Frank to indicate I was going to fire. It would make a hell of a noise inside this tent. I aimed above the *shadow* and pulled the trigger. The noise gave such a scare to our would-be intruder that he fell over backwards and down the slight incline away from the tent. I was out of the cot, over to the "door," unzipped the bottom enough to look out without being seen from outside. A half-clothed Indian scrambled away from the tent, half running, half stumbling. It was an easy shot if I wanted to hit him. Instead, I fired two rounds above his head. He stumbled, got up, ran faster, and was out of sight. Silence enveloped the camp area

Donald Haack

once again but voices could be heard. The shots had awakened even the soundest sleepers. We would explain in the morning.

"Frank, I think we scared the shit out of that guy, and I don't think he'll be back tonight anymore. We can either take turns staying awake or take my suggestion and get some much-needed sleep."

"Let's do the sleep thing. Don't think he'll be around tonight or any night."

The next morning we sorted out what really happened. It turned out our camp cook, Maria, besides cooking and cleaning was hustling a bit on the side. And our would-be intruder was really a lover-boy who was attracted to Maria—whom I had almost forgotten was taking refuge in our tent..

"Frank, Maria has to go."

20. THE VENEZUELAN TRAGEDY

Just before I left for the first Diamond Safari, my brother called to tell me that a lawsuit we were involved in and had been going on for four years was about to be settled. That seemed like another world. It had begun five years ago when Jan and I were living in Marquis, our interior home on the Brazilian/B.G. border. The U. S. Consul in Georgetown contacted the Guiana government about a downed American airplane.

The last radio contact with the plane described a crash-landing in Wai Wai country, an area between our home and the Amazon River. In *Bush Pilot in Diamond Country* there is a detailed description of the rescue. There were four persons on the aircraft when it went down: the owner of the Helio Courier; Tom Slick of Slick Oil Company of Houston, Texas; Leonard Clark, the author; Dick Macklin of Macklin Engineering in California; and the pilot, Jack Barber. After landing in the cleared area near where their plane had crash landed, I discovered the makeshift runway was much too short to take off. In a week with the help of the Wai Wai Indians, we lengthened it enough to fly out one person at a time.

Tom Slick overnighted at our home in Marquis. After dinner, he pointed to a map on the wall and asked how I happened to have it.

"It's a map of the diamond concession my brother and I have in Venezuela on the La Paragua River. Why? Have you heard of that diamond area?"

Donald Haack

"More than just heard about it. I've been trying to obtain concessions there for quite some time. Of all the strange coincidences, you are the one who has title to them? The only reason we're in British Guiana is because we couldn't get any immediate concessions in Venezuela and we didn't want to wait six months to a year."

"Interesting, that's the same reason I'm here. My brother and I spent two months in Venezuela, helping Kennicott Copper assay a newly acquired gold concession on the Amacura River near the mouth of the Orinoco River. The area was full of gold, but alas, no diamonds. In return for our help, they obtained a diamond concession for us on La Paragua. We had to wait six months before we could actively develop it so we came here— found diamonds— and stayed. We still intend to work the Venezuelan claims."

"If I didn't see this, I wouldn't believe it. The odds of the person who rescued us from a plane crash in British Guiana owning the claims I've wanted in Venezuela has to be more than one in a million possibility. Tell me, if you haven't started working the area yet, would you be interested in a partnership? You have the claims and I could finance the capital needs." *Those words are music to anyone in the mining business.*

Tom was long overdue in Texas. His original plan was to be in B.G. three weeks. It was now two-and-a-half months, he had urgent business to attend to, and had to leave immediately. I put him in touch with my brother in Milwaukee to work out the details and flew him out the next morning.

The joint-partnership with Tom took a few months to complete and based on our experience in Guiana, I ordered the equipment from Denver Equipment Company. Bob gathered information on how to float that equipment on WWII rubber pontoon barges. Tom wanted representation on the dredge and asked if it would be okay to have Dick Macklin and Leonard Clark on the mining site. Of course we agreed.

Transportation is critical to any bush operation. Venamverse, the new company, was 105 miles southeast of Ciudad Bolivar, the nearest major town and ten miles from our La Paragua dredging operation. The road from Ciudad Bolivar was rudimentary, unpaved and closed when needed most. Transportation was undependable and at the vagaries of broken trucks, bad roads, and weather. Part of the mining plan included having our own plane to operate between our base and Ciudad Boliver.

Bob hired Ernie Argentati, a young Milwaukee pilot. They found a plane in Florida that fit the operational needs: a WW II aerial survey, single-engine Convair L13 with a big cargo door, a 300 hundred horsepower engine and the ability to carry 1800 pounds of cargo.

Bob and Ernie flew the Convair L13 down through Central America into Caracas, where it was inspected and certified to fly in the interior. The Venezuelan paperwork took two months, after which Ernie maintained a regular schedule between Ciudad Bolivar and our airstrip located at our newly built bush house and cargo depot.

A sixteen-foot Alumacraft boat with a twenty-five horsepower outboard completed the transportation needs between the storage depot at the house and the dredging site. In spite of the usual hang-ups, the dredge was assembled within weeks. The diving crew slept in tents on the bank near the dredge. Another couple of weeks were spent fine-tuning the Denver Mineral Jigs using my instructions as a guide.

The first challenge—stabilize the barge. Jigs can't separate diamonds from the gravel if they're not level. The specific gravity of diamonds, only slightly heavier than gravel, required that the jigs be fine tuned: the amount of water flow regulated by pressure and number of pulses per minute, the space between the steel marbles on the bed, the amount of gravel fed into the beds, and the timing and controlling feed-off of fine material that settled on the bottom of the jigs—all had to be coordinated to keep the gravel in suspension to allow the diamonds to settle to the bottom.

Gold and diamonds less-than two millimeters accumulated at the bottom of the jig beds. The fine gravel, sand, and diamonds had to be constantly drained to keep the jigs from filling up, which would alter the flow of water. The goal was to create a quicksand effect in the jig bed allowing the diamonds, which were heavier, to work their way down to the bottom. It required a continuous coordination of all factors to efficiently extract the diamonds from the gravel. The primary requirement—the barge and jig had to be level.

With everything on schedule, we had no premonition of the tragedy to follow. Ernie flew every other day to Ciudad Bolivar for fresh food and hardware. Coordinating with Ernie, a local Indian boat captain navigated the hour-long hazardous boat trip between our house on the Caroni River to the La Paragua tributary where our dredge and claim was located. The captain was experienced. He had to be. The Caroni River widened to as much as five miles with hundreds of treacherous passages between islands. Water levels changed daily, requiring the *Capitan* to know which channels were safe and which were not. To give Tom Slick progress reports, Macklin and Clark accompanied the captain on several trips.

Leonard Clark's OSS (which later became the CIA) World War II background and flair for dramatics set the stage for the tragedy. Since coming to camp, Clark had started another book and spent hours on his portable typewriter. Like many writers of the time, he subscribed to the belief that alcohol improved writing. The cloak-and-dagger life never left Leonard and he continued to create plots and schemes. Some were real, many were not.

Intrigue, instilled during the war, was left behind by most of the OSS people when the war ended, but Leonard continued nurturing it in his life and in his new book. Unfortunately, made-up spy stories overlapped in real life with friends and associates.

Case in point: his detailed story backed up by phony papers he gave to Tom about the sabotage of our plane in Guiana by a ring of Russian diamond smugglers—detailed in *Bush Pilot.* It disrupted our lives and strained a close friendship before we were able to expose Clark's rendition of an assassination plot against me for what it was—pure fiction.

His last published book had come several years earlier and he wanted to make a splash with this new one. He came up with the idea of finding Jimmy Angel's mother-lode of gold, which Angel purportedly found after he crash-landed and abandoned his plane on top of the ten-thousand foot escarpment near Angel Falls, the tallest known waterfalls in the world. Clark's scheme was to have Ernie fly him up there and drop him off at the top to search for Jimmy Angel's gold lode. In the story, bad weather would prevent Ernie from rescuing Clark stranded above Angel Falls. The world would then be informed of the crises: Leonard Clark, adventurer, explorer, author was trapped above Angel Falls with rations running out, his chance of survival diminishing daily. This would produce the needed publicity for his new book. He tried to convince Ernie to land him on top of the plateau.

Ernie, fortunately, was both practical and sober. He thought it best to inform my brother of the hair-brained scheme. Ernie wanted to know if Clark had authority to order the plane on such a crazy mission? It had absolutely nothing to do with the mining operation. And with that thought, Ernie felt justified in stalling Clark while he contacted Bob. Clark's idea was immediately and categorically rejected along with any other crazy ideas that might be conjured up by him.

Clark was upset. After starting the day in a drinking session, he announced that they hadn't been to the mine site in several days. He convinced Dick Macklin and Ernie, who wasn't flying that day, they could take the boat to the mine site even though it was the captain's day off. Clark assured them it would be perfectly safe because he knew the Caroni and La Paragua Rivers, having been on the trip many times with the captain. The three of them jumped into the Alumacraft boat and roared off into the upper Caroni.

They reached the site on schedule and watched the crew dredge and test the equipment. A couple of half-carat diamonds were found on the bottom of the upper jig beds, but the divers knew they were still losing some of the bigger diamonds. Testing went on, a slow process because they had to shut down the dredge, empty the jigs, and sieve the fine material for small diamonds. Macklin, Clark, and Argentati were fascinated by the process and time slipped away. When they realized the late hour, they took off, assuring the diving crew they could reach camp before dark. Clark knew a short-cut, a river channel that would save them half an hour. They took that channel.

The spare divers and workmen back at the home base camp wondered why the boat hadn't returned before dark. There were no extra sleeping accommodations at the dredge. The next morning there was no sign of the boat or the three men. The captain returned and took his dugout boat up river to find out what happened. After he heard the story from the diving crew, he feared the worst. He knew the channel Clark chose could have saved time but, at the present water level, was too treacherous. No boat could make it through the heavy rapids and whirlpools. The captain took the long trip around and came up to the big whirlpool downstream of the short-cut channel. His worst fears were realized: the Alumicraft boat was on the shore lodged between rocks. The damage was severe—as if a giant hand crushed the heavy aluminum sides. Nothing was in the boat. They searched the banks for possible survivors. They dispatched men to search the banks upstream.

The Caroni always gives up its victims. In warm tropical waters, bodies decompose quickly, bloat and float. By noon the following day, they retrieved all three bodies and brought them back to camp. Bob made the decision to bury them in the banks of the Caroni River.

The following day I received a call from Bob, informing me of the tragedy. Could I come immediately to help out and fly the plane? I had little choice. I knew I had to go if for no other reason than to give him moral support.

When I arrived in Ciudad Bolivar, I found Rene Borja, the owner of the small charter planes, whom I met on earlier trips. He expressed his sympathy, as did others. Bad news travels fast. Rene dropped me off at the campsite in his Cessna 180. Bob met us at the end of the strip where the Convair L13 was off to one side. In the next hour, I heard all the details, saw the gravesites, and sat with Bob, planning on what had to be done. He had notified the American Consul in Caracas, who in turn notified the Milwaukee office and the next-of-kin. Ernie was by far the saddest case— married a few years, his daughter had just celebrated her second birthday.

The next priority was the salvage of the mining company. The boat supplying the dredge site was destroyed, and the airplane bringing supplies from Ciudad Bolivar sat on the airstrip, without a pilot. Logistically nothing moved. In less than a week the operation ground to a halt, and the dredge shut down. One man was kept on as caretaker. Three of the diving crew stayed on at the house. Bob's days were fully occupied trying to keep up with the red tape.

The American Consul required endless forms filled out. The Venezuelan government submitted just as many forms. The families wanted the remains shipped back to the States, but Venezuela's restrictions against opening graves and transporting bodies created an obstacle. It was my job to get the Convair L-13 flying again. An order for a new boat and engine was placed. Then, the cash flow of the company dried up.

After losing his two friends, Tom Slick was disinclined to invest more in the operation; but faced with losing everything, he wired enough cash to keep the company afloat. He would try to find another investor.

Meanwhile, I faced the red tape of obtaining a temporary license to fly the L 13. Our company could get by with a part-time pilot. Rene Borja, the owner of the air charter service in Ciudad Bolivar understood our predicament and offered to fly the Convair L-13 until we found another pilot. Up to this day, I don't know whether that was a ploy to monopolize the business or a real offer of help, but his actions stalled the operation and eventually broke the back of Venamverse Mining Company.

After a couple of days in Ciudad Bolivar organizing supplies, I met with Rene and flew to our camp and went through the operating manual of the L 13. He brought another pilot to fly his plane back. Rene and I flew our L13 out. Since I had more experience with short-field operations, Rene asked me to fly with him until he felt comfortable with the Convair. I agreed and suggested we practice empty and fully-loaded landings, then slow-flight stalls at 5,000 feet altitude, leaving plenty of air to recover if anything went wrong.

Rene put me off several times, saying he was busy. That should have set off the warning bells. The day set for flying into our camp was getting close and we hadn't checked out the aircraft. My only thought was that we would do it the same day before loading cargo. Bob needed steel reinforcing rods to stabilize the rubber pontoons that floated the dredge. When I arrived at the airport, Rene was already there, the plane was loaded with the steel. He was checking his watch.

"You're not going to do any short-field landings or check out how this bird stalls?' I asked rather incredulously.

"Don't have time for that. Have to be back in a couple of hours for some charters. We can go in and out quickly. This isn't a full load. You can fly right seat and I'll fly it in. Your strip isn't that short, and we shouldn't have a problem."

It's funny... no pathetic, when all the warning bells go off, yet the urgency of the situation overrides common sense. I knew Bob needed the equipment. After a week of delay, there was this sudden rush. I reluctantly and mistakenly agreed to fly with Rene.

As we approached the airstrip, I thought he was coming in a low and slow pass to see if the strip was clear. Our airspeed dropped, he put on half flaps, and we kept coming in.

"You're going straight in?" I shouted over the engine roar. "Aren't you going to do a fly-over first?" He shook his head. He was coming straight in. "Without full flaps?" I again shouted. This time he didn't answer. He was intent on the landing. He didn't do too badly, considering he only used half flaps, except we passed over the first part of the runway 10 knots too fast without touching down. He realized our bad situation too late, and without other options, pulled back power and pancaked into a jarring landing on the last half of the strip.

We stopped a few feet past the cleared area and were in the rough. A very hard landing for a fragile tail wheel. We slowed down fast after we hit. The tail wheel had broken and dragged in the dirt. Acting as an anchor it was effective, but not exactly a proper use of equipment. I jumped out, glad to be alive. I spotted our tail wheel halfway down the runway, followed by a deep groove in the grass right up to our plane— the last four feet of the tail section, askew, lay flat on the ground. This plane would not fly again soon.

It's a question whether Rene was overworked, stupid, or actually planned to put the Convair out of business so he could have all the air shuttling himself.

The plane never flew again. Venamverse ran out of funds, so Rene did not profit by a mining company being in the area.

This time there were no more operating funds from Tom Slick. He shut the operation down and paid a couple of men to watch over the house and dredge.

"I've invested more than I anticipated. I know we have good concessions but we need someone else with deep pockets, who's willing to go all the way with this operation. I have a couple of ideas."

Within a month he found a large dredge that wasn't operating— its owner was a guy named Sladder, who had family connections in the White House. Sladder's dredge had been idle for over a year. Using it on our concession appeared to be a good solution and a win-win for everyone. For contributing the dredge and bringing it to the concessions, Sladder would receive a one-third share in the company.

While this was going on, we were introduced to Tom Slick's other interests. He was fascinated by the unusual. One of his interests was *Big Foot*, better known as the *Abdominal Snowman* of the Himalayas. That's where he met Leonard Clark who was documenting his adventures in *Life in the Himalayas* after his *Wanderer 'til I die.* Tom was fascinated with the South American Indians, who used ESP (Extra Sensory Perception) and mental telepathy and questioned me in detail about experiences I'd had with them.

Tom brought a Dutchman, Peter Hurkus, who had the ability to read a person's history by holding a personal item, like a watch, ring, or other jewelry. He had been on several national TV shows and was a moderate success. Peter explained that he had this unusual ability but it wasn't always on call, which resulted in many people labeling him a fake. That bothered Herkus to the point of not wanting any more TV appearances. The problem was that he was paid well for those appearances. Each one brought more than he'd made in a year as a house painter in Holland. His life-style expanded exponentially, so he continued to do the shows in order to sustain it.

He settled in Brookfield, Wisconsin, a mile from our house in Elm Grove, so we saw him several times between shows. The first time we met was at the Milwaukee Athletic Club where my brother invited Mother and Dad. It was a pleasant dinner and conversation. Peter related how he came by this unusual talent. He fell off a ladder, struck his head, and was unconscious for several days. When he awoke, he had severe headaches. When they subsided, he noticed that when he was holding a person's keys or watch, he had flash visions of their life. When it first happened, it scared his friends and they avoided him. When the word got around what he was doing, several promoters exploited his talents. Then he met Tom Slick who brought him to the States.

Halfway through the meal, Peter reached over and asked Mother if he could hold her bracelet. She removed it and gave it to him. He held it while the conversations continued around the room. The conversations stopped when Peter spoke. He was holding Mother's bracelet in his hand and looking at it while he gave a long and deep explanation of her younger life. Mother was spellbound and dumbfounded, and as she listened there were smiles, frowns and tears as Peter went on. Some of it I knew: the large family mother came from; her French, skinny, bearded father who should have been a priest; her very heavy mother of German descent; her sister Julia, who died very young and who was close to mother. His relating to Julia was when Mother teared up. He went on to other details that Mother said she had all but forgotten. Peter's unusual ability hit a home run that night. We were all believers.

Donald Haack

Three months after Sladder signed the contract with us, Bob cabled me that not only did Sladder default on the terms, but could not be found. Tom Slick used all his connections trying to find him, to no avail. A month later Sladder showed up in New York and was less than apologetic to Tom, to the point that he said he couldn't care less and disappeared again.

Tom hired the New York law firm, Bondy & Schloss, to sue Sladder for breach of contract. The suit extended over a year. Bob and I flew to New York several times to give depositions that seemed to go on forever. In the end, Tom won the suit and a sizeable judgment, but discovered later there was no cause for celebration. When Tom attempted to seize assets, bank accounts, or real estate, he quickly discovered that Sladder was sue-proof—no visible assets. No assets were found, but the judgment was active. Months later Sladder approached Tom and offered a settlement of a few cents on the dollar—to get rid of the judgments. Tom accepted. So ended another not-so-successful quest for diamonds. And perhaps another chapter of why diamonds are so expensive.

Donald Haack

21. MEETING DANIEL LUDWIG

I was in and out of the mine site, but spent a good bit of time at the Hotel Ciudad Bolivar while working on the permits to fly the Convair L13. The hotel bar was the meeting place for all mining-related people in the district. The evening pre-dinner drinking groups provided an interesting and lively discussion time.

One evening I sat at the bar next to a quiet, well dressed man, different from the usual mining patron. He introduced himself as *Mr. Ross*. That, in itself was different. First names were the norm. Last names were rarely used. We went through the usual: "what are you doing here—where did you come from—how long are you staying?" routine.

Ross was from New York. He worked for National Bulk Carriers. Much later in the conversation, he explained that the owner was Daniel Ludwig, listed by Forbes Magazine as the richest man in America.

Ross's visit was not a pleasant one. "Mr. Ludwig owns the largest ranch in Venezuela, La Gravenia, located southwest of Ciudad Bolivar. Two days ago, the company plane took off early in the morning with the ranch manager, a man from the New York office, and two of the ranch hands. Inexplicably just after take off, apparently without any engine problems, the plane veered to the left and crashed into the ground. Ross was there to investigate the crash and then later to work out arrangements with the respective families. No one could understand why the accident happened. The company had a very competent and experienced pilot, and the plane was kept up to the highest standards."

The usual warning bells started ringing. This sounded all too familiar. "By any chance was the take off in the dark before daylight?" I asked.

He said, "Yes, they were going to meet a connecting flight in Ciudad Bolivar that left very early. They didn't want to call it too close."

"Was the take off from the ranch, and did the airstrip have lights surrounding it? He nodded affirmatively.

"Mr. Ross, I don't want to oversimplify, but this was a common problem in World War II, when our planes were taking off from remote airstrips at night. The fields were lit with oil pots along the airstrips. The Air Force only lost a few planes before they discovered the problem—and it was easy to correct." I explained in laymen's terms what happens when a pilot is on a visual take off down a lighted runway. After take off in civilized areas, there are lights all around for visual horizon reference; but once airborne in remote areas, there is total darkness. It takes a few seconds for the pilot to adjust between outside visual and the instruments inside the plane—often not enough time for the pilot to realize he is not on a straight-ahead climb out but on a high speed left-hand turn. The result? He crashes into the ground.

Ross was quiet as he swirled his scotch and two ice cubes round and round. I said nothing more and waited.

"You know, that's the logical explanation. Now that you mention it, I do remember some articles written about that problem. How did they finally solve it?"

"Once they knew the problem, there was a simple solution: The pilot stayed on visual as he took off into the darkened night. The copilot never looked out the cockpit but instead was on instruments from the beginning of the take-off run. The instant the pilot lifted off, the copilot took over. Being completely on instruments, the copilot had no problem maintaining heading and altitude. No more crashes. It cost a few airplanes and lives before they came up with that simple answer, though."

Donald Haack

Mr. Ross, I felt, had something in mind and while he was pondering it, he answered the question I had asked earlier—how did Daniel Ludwig decide to have a ranch here in the state of Ciudad Bolivar?

"You've heard of Cerro Bolivar, that mountain, almost pure iron ore, sticking up in the middle of nowhere? U.S. Steel has a company here, Orinoco Mining Company, which is literally taking down the whole mountain with bulldozers and trucks—about four times the size of anything ever seen before. The trucks haul the ore out to Porto Ordaz, at the junction of the Caroni and the Orinoco, where they have a huge bulk-transfer facility to load the ore into ocean-going ships. That's where Ludwig comes in. He's got the ships and the contract to haul the ore up to the States for refining into steel. Ludwig's company, National Bulk Carriers, is one of the biggest in the world. After loading up, the ships navigate down the Orinoco to the Atlantic. You should have seen the work they did to dredge out the channel of the Orinoco to handle those ocean-going cargo ships."

"That explains the humongous ships we saw when we were on the Orinoco in a sixty foot dugout. We almost swamped every time one of those buggers passed us, at least the ones going east fully loaded. Seemed like they moved half the Orinoco when they passed," I added.

"You were on that section of the Orinoco in a dugout?" He was incredulous. "I've seen the amount of water they push aside, and I've seen big boats get swamped as one of the ships passed. To help the situation, the company gives out timetables and uses their foghorn that can be heard twenty miles away, but still there is trouble. The Indians just don't bother with the notices and the foghorns—with disastrous results. How in the world you navigated that area in a dugout is a minor miracle."

I explained it was bigger than a usual dugout—65 feet long, but we still had our hands full when those ships passed. We encountered five of them on our journey to the mouth of the Orinoco, where we went up the Amacura River. Then he wanted to know what we were doing and where we went. Spending most of his time in a New

York office, this was fascinating to him. That night we had dinner together.

He held an envelope and a piece of paper as we sat down. "Can I ask a favor? You're going to be in New York tomorrow. Could you be so kind as to meet Mr. Ludwig and tell him the same story you told me about why the plane crashed? He's taking it quite hard that four of his people were killed and there seemed to be no plausible explanation. He would really appreciate it." He held the envelope as he waited for my answer.

"Yes, of course. I'd be glad to do that." The letter was addressed to a Mrs. Jinacki. The address was National Bulk Carriers, 1345 Avenue of the Americas, 34th floor. "This won't be any imposition. I'm going to be in the diamond district on 47th Street, a block away."

"Ask for Mrs. Jinacki and give her this letter. She's Ludwig's personal secretary and no one gets to see D.K. Ludwig unless Mrs. J okays it. If by chance she should be out, ask for Miss Young, she's blond, wears glasses, is beautiful, and works with Jinacki._She'll get you in, too. I appreciate what you're doing and don't want you to have to waste time waiting. I took the envelope and the paper with the addresses and names.

The next morning I was on my way to New York, and Ross was taking a truck to the La Gravenia ranch. My time on 47th Street took less than an hour. The diamonds I had shipped to the cutter a week earlier had been examined by him. He specialized in cutting and polishing one and two-carat stones of a certain fine quality, which was what I had for him, so it took only a matter of minutes to agree on a price for the parcel. I walked the block-and-a-half to 1345 Avenue of Americas and punched in floor 34. I was met by a receptionist who steered me to Mrs. Jinacki's office.

"Mr. Haack, we've been expecting you. Thank you for taking the time to come here."

I wondered how in the world Ross was able to get a message up to this office so quickly. But then, if you're working for the richest man in America, I suppose he has sophisticated communication systems.

Donald Haack

Mrs. Jinacki was certainly the right person to handle personnel; she was warm, friendly, and made me feel like a very important person. "Mr. Ludwig is on the phone but knows you're here and will see you in a few minutes. Tell me about what you do in South America. Mr. Ross said you had an interesting background—you're a pilot, have your own plane, and you buy and sell rough diamonds. How fascinating! I want to hear about it." But before I could answer, the door behind her opened and Daniel K. Ludwig came in, grabbed my hand, and ushered me back into his office, which was big but not ostentatious. He had a good view of the city.

He wanted to know all the details I had told Ross and more. I thought that was the end of the conversation, but he felt like talking and explained how he got into the shipping business. A good buy on his first cargo boat, financed it up to the hilt while he took on any and every shipping job that he could find. Soon he was on his second ship and the two ships financed the third and so on. He explained how big his company was and how many ships, but the numbers overwhelmed me. I stayed much longer than I expected, and when I finally departed, he made me promise to come back again and bring some rough diamonds. He had never seen one.

Eight months later, when I had a parcel of diamonds to sell to Anton Smit, the industrial house, and Harry Winston, I called Ludwig and he invited me to come right over. I offered to take him to lunch. "Just come right over, the timing is right."

When I got to his office, Mrs. Jinacki guided me right into his office.

"Good, I'm glad you could come. I have my lunch right here and there's enough for both of us. Never go out for lunch. Takes too much time, and that's the trouble with most people nowadays. They want to know how much vacation they can take, how many holidays and days off. They spend a lot of time at the water coolers, take long lunch hours. Didn't build this business doing that. Worked long hours…sometimes all weekend. That room over there," and he pointed to an adjoining room, "has a bed, shower, and anything I might need when I sleep here. Get a lot of work done if I don't have to spend that time commuting. That's what's wrong with people today…they don't want to work. Here's a sandwich, some carrots, and an apple. If you're still hungry there's more where that came from. Let's see those diamonds now."

Between bites I spread out the 150 carats of diamonds, keeping the two parcels for Winston and Anton Smit separate, and handed him the tweezers and my diamond loupe if he wanted to examine the rough diamonds. He did, and I had to show him how to hold the diamonds and how close to position the loupe, so he could see inside the diamonds with the 10-power magnification. He really enjoyed doing it, so much so that he called in Mrs. Jinacki, Miss Young, and finally half the staff in his office, as he told them the stories on how the diamonds were found.

Later he confided to me that he was developing a huge project on the Amazon River, the biggest ever conceived. I knew he didn't do things on a small scale, but when I heard the details, even I was surprised at the immense scope of the project. His plan was to harvest timber inland along the Amazon River. Because the area had so diverse a quality of hardwoods, it was not cost-efficient to pick out all of one kind, which is what had to be done to get a proper selling market.

He contracted to build a complete floating city, with processing plants that could handle all the hardwoods, turning them into pulp and then plywood sheets for which there was a world market. Simultaneously, he would replant the clear-cut jungle with specific species of hardwoods commanding the highest prices. At the equator, with its twelve-months growing season, the newly-planted trees would mature in half the time it would take anywhere else. It was indeed a grand plan.

The city and processing plants were being built in Kobe, Japan, on the biggest barge/ship ever built. The plant was capable of being towed across the world into the Amazon and anchored there, until they had harvested the trees in the area. Then it would move on to the next harvest site. The concept was mind-boggling.

In retrospect, I shouldn't have made any comment, but having heard so many horror stories about partnerships in Brazil that didn't work out because of lax laws and corrupt partners, I brought up this matter. He was about to spend over a hundred- million on this project, the biggest he ever tackled. Brazil did not have a good record of partnerships at the time, and I asked how he expected to overcome that potential problem. He did not like the question and skirted around it, saying, of course he checked that out thoroughly and that would not be a problem. Story done.

We ended on a good note and he extended an invitation to stop by when next in New York. He sounded like he was truly interested in what I was doing. I promised to call him or stop by on my next trip but it turned out to be a long time.

Two years later while again in the big city, I called him from my friend's office. The response was lukewarm, and even less so when I asked about the Brazilian timber plant on the Orinoco. He cut me off short and said he was busy. I did not get an invitation to stop by.

A year later I read in the Wall Street Journal about the disastrous business project that Daniel K. Ludwig, owner of National Bulk carriers and formerly the richest man in America experienced in trying to work with an irascible and corrupt Brazilian government. In short, Ludwig was forced out and lost most of his investment.

Thirty-five years later, I met another of Forbes' richest men in America, Bill Gates. He spoke at Queen's College in Charlotte, NC, and Jan and I were invited to his talk and later to meet him at a special reception. We talked about South America. Bill said one of the most fun and adventurous trips he had was a boat trip up the Amazon. We determined he was about 400 miles south of our one-time home on the Mahu River, a tributary of the Amazon.

I couldn't help but muse: the two richest men in America both fascinated by South America and the Amazon River. Ludwig invested a large portion of his wealth in that continent, whereas Bill Gates enjoyed the adventure but wisely stayed away from any investment.

22. THE ACCIDENT

The Safaris were finished for the season. The flying was regular but not excessive. Frank supervised the dredge and had a good supply of rations and fuel to last him awhile. Georgetown had its share of social gatherings led by the very outgoing party group of American shrimping companies based in town.

I was living in "Mama" Mittlehauser's boarding house where all the Gorinsky children stayed during their school years. With no proper schooling in the interior, the children left home at the age of five and Mama raised them until "O" level's (our high school) graduation. The children were only back at the ranch for Christmas, Easter, and summer holidays. Mama was a real taskmaster, so Caesar and Nellie had no worries that their children were properly raised and disciplined.

Caesar suggested I stay there, as he did on most of his trips to town. I had a quiet and peaceful room, exceptional breakfasts, and home-cooked meals when I wanted them. Mama looked after me like a mother hen. When I returned home one night, Mama was waiting at the door and handed me a fancy invitation from the shrimping company manager, whom I had met at an earlier fete. Mama patiently waited for me to open it and tell her what the invitation was all about. With all the Gorinsky children gone, now, Mama directed her motherly attention to me, sometimes to a fault. I showed her the invitation. She read it and said, "You stay home too much. You should get out more often. Tell them you'll go. Probably meet some new friends." She handed it back to me. "It's RSVP. Call them now," as she pointed to the phone. I did what she bid— and accepted the invitation. "And wear a suit and tie," she added. "It's in the nicest part of town. I know the house. I cater for them sometimes."

That night my car was still in the garage for repairs, so I called a taxi. I arrived a bit late; the big crowd—the people who owned and ran the town—had already gathered. I stepped inside, as I was handed a drink and met the American Consulate and his two department heads. They seemed sincerely happy to meet a fellow American. There weren't many in town.

Georgetown parties did not lack drinks. They flowed like the Demerara River. The best drink in the country was *Demerara Rum,* and it was everywhere. There wasn't much to do in Georgetown, which encouraged serious drinking. On Sundays, people went to church, after which they gathered for a few drinks. Sunday afternoons, families walked along the sea wall where a band played in the gazebo. The Demerara and Essequiba Rivers carried soil from the interior and dumped it into the sea, so mud flats took the place of sandy beaches.

It was, as most parties, formal and proper at the start and usually raucous toward the end, which was anytime between 12 midnight and 4 a.m. Midway into the evening a couple of the livelier inebriated guests started jumping in or pushing guests into the pool, where they squealed and splashed anyone nearby. I moved to the other side of the area, closer to the protection of the bar. I didn't want to appear like a party pooper and leave earlier, but at one o'clock I bid my host and hostess goodnight. John, the American Vice counsel was standing alongside with girl in hand. "I've had my share to drink and we're going, too. Heard you wanted a taxi. Don't bother. Going to Brickdam and I'll drop you off on the way. Getting a taxi this time of night could be a long time coming." Gratefully, I accepted his offer.

John was not too steady. I offered to drive. He declined.

"No, no, I'm fine. Hop in. The back seat is full of Consulate junk. Get in the front. There's plenty of room."

John got in the driver's side. I helped his girl friend, Christine, into the front middle seat and followed on the right seat. John was feeling no pain. It was a relief to see he wasn't speeding. In fact, we inched along. We came to a complete halt at the stop sign on the street before Brickdam, but since John was in slow motion mode, we sat there for a couple of seconds. I was about to make a joke of it, when I saw a car speeding down from the left and assumed that was the reason for John's lengthy stop, even though the car was a long way down the road. As the car approached, John surprisingly shot forward into the intersection and inevitably into the path of the oncoming car. It all happened in a millisecond. I saw the car and then felt the impact.

How could a car hit us with that speed with no noise? Everything was silent, yet I was sure the speeding car "T-boned" our car. I was confused. There was no noise, in fact it was eerily quiet. My neck was stiff. I turned slowly towards John and Christine, who were slumped over, unconscious. Something was in my mouth. I tried to lift my left arm. It didn't respond. I reached up with my right hand and unsteadily touched my mouth. My tongue pushed out the object, and I held it between my thumb and forefinger, everything still in slow motion. I held the object up and moved it around until some light fell on it. It was my tooth. Slowly the perception of what happened became apparent. The crash must have been worse than I thought. I didn't hear any sound because I was unconscious for a few moments. I couldn't see anything out of the windows, the front windshield was partially smashed, but it was dark on both sides. We must be down in a ditch between the two streets. There was a moan from my left. John was moving slightly. I called out, "John, are you okay?" No answer. Christine was not moving. I leaned over to check on John. There was an excruciating pain in my chest. What I saw was not good. John's forehead was caved in and blood was coming down both sides of his face. I knew I had to get out and get some help immediately.

Donald Haack

I pushed and kicked on my door. It didn't move no matter how hard I tried. The handle worked, but the door had been jammed shut. I couldn't get across Christine and John. My only exit would be through the back seat. I inched up. That's when I realized there were more serious injuries. The pain on my forehead was from a bleeding cut. When I touched it, I could feel glass shards. Fortunately, the wound wasn't deep and the bleeding was not serious. I had other pressing problems. My left arm was almost paralyzed—my elbow was numb and the fingers hurt and were slow to move. The pains in my chest, I knew, were from broken or badly bruised ribs. My leg was another matter. It didn't seem to be part of me. Any movement shot pains up the leg and into the hip.

I pushed, pulled, using my right leg and arm as fulcrums to slide up and over into the back seat. At that point everything hurt, but my goal was to get out of the car. The door didn't open when I turned the handle. I lay down on the back seat and, using my right leg, kicked with all my might against the door. It popped open. It took minutes before I could extricate myself off the seat and through the half-open door. The door couldn't open fully because it was against a steep bank. When we were hit we must have careened down the road beyond the crossroad and into the ditch between the two parallel roads. If this were rainy season, the ditch would be flooded and we would be statistics. I couldn't move on my stomach or side. I turned onto my back and with one good arm and one leg, hitched up the bank inch by inch. I reached the top. No one was around. I called out,

"Help, there's been an accident. We need an ambulance." No answer. After a couple of calls someone came running up. "Call an..." I felt his hands pulling open my suit and trying to get my wallet. I swung at him with my good arm and yelled out. "Help, there's a thief here. Police!" I got one good swing at him again, as he tried to get my diamond stick pin. The punch landed on his nose and he drew back. Something got his attention, and he got up and ran as a man and a woman came up. "Thank God you're here. That guy who ran away was trying to steal my wallet instead of calling an ambulance."

Donald Haack

"It's okay. He's gone now. We live across the street and heard this loud crash. We called an ambulance and came as quickly as possible. That ghoul trying to rob you ran as soon as we came. Where's the car you were in?"

"Down in the ditch," and I pointed with my right hand.

"Oh, my God! It's really smashed," I heard him say, as an ambulance pulled up. They waved it over to where we were.

"Get the people out of the front seat," I directed. "It's an American Embassy car and the Vice Consul and his girl are in pretty bad shape. They weren't conscious. Take them first if there's not room for three. This couple will wait with me until you return or send another ambulance. You *can* stay, right?" I asked.

"Of course we'll stay with you."

"Are you sure you're okay?" the ambulance driver asked.

"Yeah, as long as I don't move. I've got a couple of broken bones, but I don't think I'm bleeding, and I'm already lying down if I go into shock." The other two from the ambulance were working to open the car door. They went back for tools. The driver returned with a blanket and handed it to the woman.

"He seems to know his symptoms. If he does go into shock, cover him with this and elevate his feet. That should work until we get back in fifteen minutes—if we can get these other two out of the car. It's a mess." Off he went

Before the ambulance returned, I asked the couple to call Mrs. Mittlehauser to tell her what happened, and that I would be in the public hospital if she wants to come in the morning.

The ambulance returned quickly, and in a matter of minutes I was wheeled into the hospital on a gurney and into the *Seaman's Ward* on the second floor. I was unattended on the gurney, and no one came to check on me. I lay there until morning, when I was discovered by the nurse on the new shift.

"You mean they left you here all night and no one came by to check up on you?" she asked incredulously.

"Well, I made it through the night, didn't sleep, but I really have to take a pee and would hate to do it all over this gurney. Can you get me something…and quickly?" She returned in less than a minute, and I relieved myself. It wasn't easy with just one arm and one leg to maneuver. It was the first relief since the accident, but I was also becoming aware of my injuries. They were more extensive than I first thought. The shock that goes with and temporarily protects the body wore off, and I was getting the full-blown effect of whatever was wrong with me. I had an idea it was plenty.

The nurse was very upset that after being in an accident no one had checked on me all night. She cleaned me up and notified the first available doctor to examine me. I had dried blood over my face, and she started there to remove as many shards of glass as she could with tweezers. She put in the order for the doctor and finished cleaning me up as Mama came storming in.

Nobody ignored Mama when she had a mind-set or was angry. She was very vocal in telling the nurse that she couldn't find out if I was even in the hospital when she first called early in the morning and continued until now, 11:30 a.m. She hopped on her bicycle and came directly to the hospital, by-passing all the usual protocol after she discovered I was in the Seaman's Ward.

"The doctor will be here in five minutes," she stated with authority. "They didn't even know you were in the hospital until a few minutes ago. How are you? What happened? I heard the accident two blocks away—it woke me up. Had no idea you might be in that car. Saw it coming over. Pretty bad. Who else was in there with you? Are they okay, alive?"

My head was spinning from shock, lack of sleep, and too many questions at one time. I tried to explain what happened as much as I could recall. "Could you ask around and find out what happened to the other two, John, the Vice consul and Christine, his girl? They were pretty bad when I saw them last."

Mama returned in a half hour. "They brought a plane in from Panama early this morning, and they took the Vice Consul up there. He has a concussion, a crushed-in forehead, and several broken bones. Too serious to have treatment here, so he will be staying at the military hospital in Panama. They think he'll pull through. Christine was critical when they brought her in last night.

Maybe that's why they ignored you until this morning. Her spleen ruptured, she has a broken pelvis, broken arm, and a partially torn bladder. They operated on her right after admission, and she is in recovery now. No news on how she's doing, but she's in intensive care with two nurses. They expect her to regain consciousness in an hour or two at most. The doctor is going to examine you in a few minutes to see how extensive your damages are."

Leave it to Mama. She was short, to the point, and no-nonsense. It was good to have someone like her around, especially at a time and place like this—not exactly a shining example of efficiency. She asked if I had breakfast. I told her no. I didn't want anything but liquids.

"You have to eat. Why only liquids?"

"Mama, if you have to know, there is something drastically wrong with my left leg and bottom half, and the last thing I want to do is have to go to the bathroom or even contortion myself to get on a bed pan. No food in, no food out, at least until I can figure out what has to be done and how to do it." She frowned but had no retort.

The examination and X-rays took a couple of hours, and I was dumped back in the Seaman's Ward to stare at the whitewashed walls. No one else was in this large room, not even a nurse. Mama came back before dark with a bowl of soup, which I gulped down ravenously and thanked her profusely. The hospital brought a tray in earlier that looked bad, smelled bad, and tasted worse. It was on the adjoining gurney, untouched.

"I called Jan and explained all I could. She insists on coming here, but I told her to wait until I heard from Mr. Lee, your doctor. He should be here the first thing in the morning. I want to be here when he comes to hear what he says, so I can tell Jan first hand."

"*Mister* Lee?" I questioned. "Isn't he a doctor?"

"In Britain, after a doctor becomes a specialist, he goes by the title of Mister. It's a grade above being a doctor. Mr. Lee is the only one with that title in the hospital. At least you have the best looking after you. Must be the embassy had a hand in finally getting you some preferential treatment, but you could have died the first six hours they brought you in and no one would have known."

Mr. Lee, the Chinese specialist, was in my room almost at the crack of dawn, followed immediately by Mama with more liquid breakfast.

"We've examined the x-rays. You have multiple fractures of the pelvis. The better news is that everything is in place, and we won't have to operate to pin them together. It's a bit unusual, because the fractures go all the way through, but for whatever reason, they are all in place, so there is no reason to disturb them, and they should heal in place…if you don't put any unusual stress on the left leg. There are two broken ribs on the left and one on the right…also in place and should be left to heal on their own. They'll be painful, but shouldn't give you any permanent problem. Your left elbow and thumb are sprained, and you've got contusions on the forehead. The glass is removed and the dressing is fine. I don't have to tell you that you have a split lip and a front tooth broken off—a good pirate image if I ever saw one. No critical damage that needs surgery.

In short, you've got a long bed rest coming up as your pelvis mends. You won't be walking for three months. With a good recovery, you'll be using a cane for several more weeks after that. You're in for a long rest."

The prognosis wasn't good, but not far from what I expected. The saving grace—the pelvis was in place. The down side—the time I'd be out of circulation. "Mr. Lee, if I could get to Grenada …could it be arranged that I check out of here at any time? I believe I can heal a lot quicker if I can get in the ocean and do light-to-moderate exercise without straining. What do you think?"

"One, I don't know how you could move in and out of cars and airplanes without straining yourself and damaging your pelvis. Two, even if you did, I don't know how getting in and out of the ocean would help your case or shorten your time. You're going to have to reconcile to the fact that you will be incapacitated for several months. Compared to the other two in the car, you were lucky. If they recover, they won't be back to a normal life for six to twelve months. As for the young man, he may never be back to normal. Consider yourself lucky."

I rephrased my question. He sighed. "Yes, I will release you anytime you ask, but it is certainly not my recommendation to do so." He turned and left. That was day two.

Mama called Jan that night and she showed up the following afternoon. I still looked a bit of a mess but much better than the day before. Jan was glad to see that I still had my sense of humor but was concerned over the gravity of the injuries. I told her of my plan to go to Grenada. She looked around the stark room. Mama had already explained that I would not eat and the reason for it.

"I've checked with BOAC from here to Trinidad and LIAT, the feeder airline out of Trinidad to Grenada. There's a connecting flight early tomorrow morning. We'll have to buy nine seats for you on your stretcher. They'll use the food lift to get you into the airplane and put you on the floor in the front of the cabin. Really quite helpful, and I do think you'll heal a heck of a lot faster over there than in here. Besides, you can't starve yourself much longer." *Good ol' Jan. She always comes through.*

They got the body to the airport in an ambulance where they deposited me onto the food tray and into BOAC's huge airplane. "Jan, I have an idea. Tip the stewardess $10 and tell her you and the patient are badly in need of a drink, like two double martinis." The flight attendant wouldn't accept the $10, but she brought back two frosty, frozen martinis. The best I've ever had–bar none.

In Trinidad, I went from the food tray to the baggage truck and back onto Liat's food tray, a maneuver that wasn't easy. LIAT was a smaller airline, but with ample pushing, pulling, lifting, and shouting they finally got me stowed on the floor next to the food. The double martinis relaxed me enough so I didn't notice the inefficiencies.

It was a short trip to Grenada. Brian Thomas, the owner of the Calabash Hotel, was waiting on the ramp with a local ambulance. It was an old beat-up one, but it succeeded in bringing us to our house. Two of the neighbors were waiting and carried me from the ambulance onto a mattress on several Bank Brewery Beer cases in the dining room. That was to be my bedroom for the next several weeks.

Everyone in our L'anse Aux Epines Bay came to visit the patient. I had a plan in mind. I asked Jan to get me a stretcher, an inner tube with a rope wrapped around it, six cold bottles of Banks Beer, and a small piece of plywood.

The stage was set the next morning. The most difficult maneuver was taking off my underwear and putting on my swimming suit, while balancing on the Banks Beer crates. As requested, six of our neighbors showed up, carried me Macbeth style, placed me in a pickup truck and drove me down to the water's edge of the beach. Everyone wondered what in the hell I was up to and wouldn't leave until I unfurled my plan. The manager of Banks Brewery was made privy to it when Jan asked to buy six cold beers. He scoffed and brought a whole case packed in ice and said I could have as many beers for as many days as I wanted, courtesy of Banks Brewery.

Jan and friends helped me out of the truck where I inched off the stretcher and into the water. When the water was up to my chin, I wondered if this was a good idea. I had very little ability to keep from dunking under and drowning. Though the waves in this bay were miniscule, every movement needed a counter movement to keep from tilting upside down. My right leg became the anchor and the right arm, the paddle or stabilizer. Jan floated the inner tube alongside and held it while I adapted and adjusted to this new element. She knew I was having trouble.

"Can you do anything with your left arm?" She gently lifted my left hand and opened it around the rope attached to the inner tube. "If you don't jerk your arm you can stabilize yourself until you're into deeper water where you can be upright and maneuver with your right leg. Besides, you'll need your right hand to drink the beer. I'll open two of them for you, and I'm staying right here to see how you handle this, so get used to it. I may need a beer. I can't believe we're really doing this, you know." At that point, I couldn't either. *In for a penny, in for a pound. I certainly can't back out of this now. Kind of lose face, as the Chinese would say.*

I had to reassure myself this was the same procedure I used in Cherry Point during the "Korean War." I tried to bring back the episode to help bolster my belief that I was doing the right thing.

Back then I was bored doing an office job. I had a re-occurrence of my Parris Island boot camp fungus on my feet. Something I picked up in the showers. And with constantly wet socks and heat, it incubated into a full-blown case of the "creeping crud," as it was called. Every day I went to the infirmary to soak my feet in a hot potassium permanganate solution, which left my feet a deep purple-brown color and the brunt of many jokes and questions in the barracks.

To pass the time while soaking, I read several books and watched the hospital fill up to capacity, and then overflow with marines crammed six to a room into what was supposed to be a double. Hallways were filled with makeshift cots and mattresses lined up along the walls. Recently back from Korea the marines were in various stages of recovery. Facilities were overloaded to near disaster. Bed sores were endemic. Those not able to crawl to the head were less successfully using bedpans, and without enough orderlies, it was chaos—not to mention the strain on olfactory senses.

Two of the overworked doctors with double bars on their collars were there regularly. We got to know each other the first week, and I took the opportunity to bring up the subject I had been considering. "What would you say if I told you that I think we can empty out this hospital in a month and get it back to normal numbers—getting them healed and out of the hospital? Want to give it a try?" I explained my plan. They crossed the room and talked. After several minutes, they turned back to me.

"Do you really think it will work?"

"Tell me, what do we have to lose? It won't hurt the patients, and if it works, we'll have a normal hospital in a few weeks." They decided to give it a try. Without the spare time to do it themselves, they wrote a note to my CO, Col. Fitzpatrick, asking to use my services part time in the hospital for a couple of weeks.

Donald Haack

Kirkpatrick was curious as hell as why two captains wanted me for hospital duty. He held the letter and waited. I explained. He was puzzled until I cleared up all of his questions. He sat back in his chair and chuckled.

"I'll be damned. It's the craziest thing I've heard in a long time but you seem convinced it'll work. This won't go any higher than me or it'll get stuck, but make sure no one does anything stupid that will get my ass in a sling. Go for it and good luck." He signed the paper giving me a month to work at the medical department at my discretion.

I took it back to my doctor friends with a list of all the material I needed. I gave the list of needed supplies to Captain Barringer and he read it aloud. "Thirty-five truck inner tubes, 200 feet of ½ inch rope, ten stretchers, five dollies, five strong corpsmen, and the use of the base swimming pool from 8 a.m. to 2 p.m., when it currently isn't being used. You know if anyone catches us doing this before we prove it works, they're going to send all three of us to the loony bin. Okay, here's your requisition slip. I don't have the authority to give you all of this, but I doubt if anyone will question it. If they do, our office telephone numbers are on there and our nurses know how to respond. I'll say one thing," the captain added, "it sure won't be dull around here." The die was cast.

I picked the five corpsmen from our SOS Squadron whose first duty was to inflate and hose down the tires, so we wouldn't contaminate the pool with the black powder that the inner tubes were dusted with that kept them from sticking inside the tires. In the meantime, we chose thirty-five patients who didn't have any open sores. We had the orderlies do a better-than-usual morning sponge bath cleaning them up. We didn't want to screw up by dirtying the pool. I made arrangements for the pool maintenance people to increase the amount of chlorine and keep the filters running a few extra hours. After explaining to each cohort what we were trying to do, my doctor friends and I were amazed at the cooperation we received. Almost everyone knew this was an end run around official rules that could otherwise take months to get approved. If it worked, we knew we were doing our best to assist the war effort and particularly supporting the guys who were on the front line. With that much enthusiasm behind it, this scheme had to work.

It didn't go without glitches. The wheels on the dollies were made only for hospital corridors and stuck when we tried a shortcut on the grass. We wound up carrying our guys to the pool. It didn't slow us down, but we had a real workout. But what the heck, we were supposed to be in good shape. And we soon were.

The next biggest challenge was lowering our patients into the pool where we helped them hold onto an inner tube and do whatever limited movement they were capable of. But getting them into the pool wasn't easy. Until they stabilized, we couldn't slide them in without help. It wouldn't be good form to have them drown during their exercises, so we had our crew walk them into the shallow end and slowly lower them in the water until they were comfortable. That caused another problem: when the corpsmen, soaking wet from the pool, returned to the sick bay for the next patient, they tracked in water, making the halls slippery and dangerous—solved by buying ten pairs of flip-flops, corpsmen wearing swim suits, and using towels to dry feet.

One of our rules was no peeing or crapping in the pool or the offender would be blacklisted in hell; no more pool and he would be assigned the worst orderlies to attend his needs. There must have been some really bad orderlies because in the three-month period we didn't have one infraction. There were a couple of close calls when patients told us *they couldn't hold it*. They were quickly lifted out of the pool in a stretcher, stood upright and peed in the grass. The corpsmen stood around, so as not to let anyone know what we were doing to the Marine Corps' landscape.

The water therapy and exercises produced minor miracles. Three or four days in the pool produced better results than weeks in sick bay—bones healed quickly. No more bed sores. Sprains and ligament problems, wounds from shrapnel and rifle holes literally disappeared in a matter of days. When the patients were ambulatory, they graduated to easy walks on level ground that kept their metabolism peaked. No longer languishing all day on hall mattress, they were ecstatic to be independent. They went to the mess hall on their own and to the head (bathroom), thereby eliminating staff who had to act as busboys and toilet nurses. Freed up of those unpleasant duties, the orderlies were more than happy to help us. It was working. The patients' appetites and spirits were up and the discharge rate was half the time I had hoped and planned.

Col. Kirkpatrick was elated that I didn't get him in trouble and was impressed by what we accomplished. I got a PFC to Corporal meritorious promotion, which included a raise that financed my fortnightly flounder and shrimp dinners at the *Sanitary Fish Market,* my favorite restaurant, in Morehead City, fifteen miles away.

With that much healing experience in water therapy, I was convinced I could do the same for myself. I had to keep that in mind while I struggled in the L'anse Aux Epines bay to stay upright in the gentle swell threatening to tip me upside down.

It took the better part of the first day to figure out how to stabilize before I tried any maneuvers. Day number one was a success. I was exhausted and slept like the dead. Day two I mastered the walking procedure. Holding on to the inner tube, I moved my right foot in small forward steps, while I made a feeble effort to move the left leg half speed with no weight on it. By the third day, I was able to gently place a miniscule bit of weight on—enough to be without pain but still have full movement. I graduated beyond the inner tube, which I anchored in shallow water and used for beer, sandwiches, and salad that Jan brought. I ate solid foods without worrying about the consequences.

Six hours of exercise a day, seven days a week, for two weeks produced miraculous results. Leg muscles were strengthened and the continued light stress on the bones encouraged a build-up of bone that healed the pelvis in record time. The ribs no longer hurt, which meant they, too, were well ahead of the healing stage. Using light swimming strokes, the sprain of the left elbow was almost back to normal—enough to hold a cane with the left arm to keep the full weight off the left leg on slow walks. I alternated water therapy with an hour of walking on the beach. At the end of the second week, I no longer used the cane and drove to and from the beach, a chore that Jan was thankful to give up. By the third week, I was walking, ready to return to B.G. and salvage our operation.

My first stop in Georgetown was to Mr. Lee's office. I told the lady who headed his department who I was, and it was urgent I see Mr. Lee if only for a few minutes. I was leaving for the interior. She sent a nurse with the message, and in a few minutes I was called into his office. His expression was as if he were seeing a ghost.

"You're walking?—without a cane? He was at a loss for more words. Then he had a torrent of questions. He made me walk back and forth.

"Is there any pain?"

"No, not in the least, except I feel a slight twinge in the groin muscle on the inside of the left leg. But that's it." He wanted to know how I got on two airplanes, home from the Grenada airport, into the ocean and what I did once in the water. He had an inquisitive mind and wanted every detail. I knew he would use some of this on his future patients.

"I know you're in a hurry to get back in the plane, but remember, no lifting. Don't push it and set yourself back. Promise me when you're in town next to come here. I want to X-ray the pelvis to see the growth pattern…and to make sure everything is lined up. Not that we would do anything. If it works, don't change it. We shook hands. He was still shaking his head in disbelief. "Thanks for stopping by. If I hadn't seen you today, I wouldn't have believed it."

Donald Haack

Don receiving meritorious promotion from Col Kirkpatrick.

Don supported by 2 chairs. Julie alongside & Diana, Tom and Todd.

23. BODY BAGS

A s an adjunct to the safaris, the charter flying was an easy, lucrative business. There was always something or someone interesting keeping the job from being dull—if you discounted those few hair-raising, adrenalin-producing close calls that flying in the bush guaranteed. What I didn't like were the requests from town to transport the remains of a family member who worked in the bush and got himself killed, shot, or drowned.

In Alaska it wouldn't be as onerous a job because cold weather preserves bodies—not so in the tropics. The Saturday night murder and mayhem victims started decomposing within 24 hours— a good reason for immediate burial. Generally it took a week or more before tragic news reached the Georgetown family.

The strong family traditions in some communities insisted that the remains be brought to Georgetown for proper burial in the family cemetery plot. Immediately after I completed my first *funeral run*, I quadrupled my fee. The experience was that bad.

The zippered plastic body bags never properly sealed. Nothing induces gagging more effectively than sitting in a closed airplane cockpit for two hours with a decomposing body. To make the situation worse, there weren't many miners in the bush offering to help place the guest of honor in a bag. The most I could expect was to pay someone handsomely to dig up the body and leave it alongside the grave. No amount of cajoling or bribing could convince a prospector to help with a stiffened-corpse-body-bag problem. That was left to my resources and it reminded me of the saying, *"How do you get a 10 pound piece of something into a 5 pound bag?"* I can't remember the answer— maybe because there wasn't one.

Sometimes gallows humor raised its ugly head. Such, I believe, was the case one particular Saturday night. The family insisted I bring the man down from Ekereku. He had been stabbed to death in a brawl two weeks earlier. I made arrangements to have the body dug up when I came on Saturday noon for pick up. This scenario was typical.

The poor bloke was stabbed outside the trading store bar on Friday, and it wasn't until Monday the storekeeper noticed him sprawled out on the grass. He dug a large shallow grave and laid the victim in—a perfectly adequate bush burial.

Rigor mortis had set in and with arms and legs spread out and it was impossible to fit him in a bag. Body bags are made for corpses whose arms are neatly folded and legs extended. The bags are 6', 6" long by 2', 6" wide, *not* 6 feet wide, which is what this sprawled-out cadaver was. No one offered to help *compact* him to fit in the bag. It was left for me to improvise the task, after which I decided my quadrupled fee was still too low.

To make matters worse, the decomposition of this particular corpse was the worst I had experienced. I flew the whole trip to town with two Kleenexes stuck up my nostrils and a handkerchief in my mouth. Breathing was touchy, and I did as little of it as possible, but it took three days before the taste disappeared from the back of my throat.

I called in to the tower to have them warn my driver, Bertie, to hire a van instead of putting this corpse in his car. It wouldn't easily fit in the trunk and the smell would linger for days. We had experienced this before. But between the Kleenex in my nose and the handkerchief in my mouth, the tower personnel evidently did not understand my instructions. Bertie was allowed to discover our problem when he drove up to the plane.

Our instructions were to leave the body at Lee's Funeral Home. We arrived there after sunset, in diminishing light. The office downstairs of the two-story building was dark. The six-foot high steel fence surrounding the office and living quarters contained a fierce-looking dog that barked furiously as we approached. There was a light on in the living quarters upstairs. Bertie rang the bell on the gate, further provoking the dog but receiving no response from anyone inside. Bertie came back to me as I was standing on the lee side of the car, breathing fresh air.

"No one's answering, Cap, now what?"

"Whoever's inside certainly could hear the bell. They're not bothering to answer. We're not leaving the cargo in the car." I looked around for small stones or pieces of wood.

"What are you doing?"

"I'm going to toss this junk at the window until someone answers or the window breaks, that's what I'm going to do." The sticks and stones against the window got bigger and bigger until finally someone raised the window and cautiously looked out.

"Who's out there and what are you doing? I'm going to call the police if you don't stop and go away."

I stared at the apparition: a knock-down beautiful Chinese lady, long dark hair, a picture perfect seductive oriental face, and a body, at least the top part that I could see, exceedingly well-endowed and barely covered in a frilly see-through nightie. It took a few moments to subdue the wild hormone attack before I blurted out, "Yes, yes, please do call the police. Maybe they can help us decide what to do with your customer here, that is, if you can't."

"Who's the customer, and why can't he speak for himself? And who are you?"

I explained who I was and why her customer couldn't speak— he was wearing a body bag. "The Rodriguez family arranged for Lee's to do the funeral and we were instructed to leave him here, which is why we are here now."

"Well, I'm sorry, Mr. Lee is out of town and we're closed. You will have to come back tomorrow. She slammed the window shut. More sticks, more stones that almost broke the window. The window came up once again. The head came out.

"I told you we're closed now. Please go away and come back tomorrow."

"Wait, don't close the window. There is something urgent I have to know. Is your dog well fed, so that after Bertie and I lift your customer over your fence, your dog won't mutilate him? The Rodriguez' family would be very distressed, and that could mean a great deal of reconstruction work for Mr. Lee— if they want an open-casket funeral, or for verification that it is the Rodriguez it is supposed to be." Bertie opened the trunk and we held the customer between us. "Shall we drop him over the gate? Or do you want him in the grass along the office?" I offered, sounding as business-like as possible.

"No, no, you can't to that. Wait a minute, I'll be right down." Seconds later the lights in the office came on. The apparition didn't bother to put on anything over her flimsy nightie. She was even more alluring full length than she was in the window. It wasn't fair. She tied up the dog and opened the gate for us to come in. She winced visibly as we passed upwind from her. She pointed to the back room, preceded us, and opened the door to a refrigerated room. "In there, please." We stood Rodriguez inside along the wall and returned to the office. Her demeanor changed completely.

"Thank you. Mr. Lee will appreciate your help in this. He is out of town." *She had already said this twice.* "This must have been difficult for you. Can I offer you some tea or a stiffer drink? It will only take a minute."

Her nightie was not tied at the waist and was partially opened as she walked to the cabinet. Her body made the classic *Varga Girl* look like a bag lady. I had never seen anything so bewitching. It made my knees weak. That basic animalistic built-in reproductive system response, which gives a strange strangulation feeling in the loins, makes the knees weak, the stomach muscles slack, the jaw hang useless, and draws whatever blood from the brain into other specific parts of the programmed procreation network of the human body—all came into play. In this case, the brain was quickly and sufficiently drained, preventing any kind of verbal answer to her offer of tea or otherwise.

Donald Haack

Fortunately, out of the mist of emotions came a voice of reason: "Cap, I think we better go. We got rid of our cargo."

Of course, how sensible. How come I didn't think of that? We left.

24. TROUBLE DOWN BELOW

The last of the tourists departed and the weather remained good, which gave us a breather. For quite some time Frank was eager to run a rig and make a strike on his own. I offered to fly him to Imbambadai to run our 6-inch diameter suction dredge, which could process 15 tons of material per hour. Even though he had a hand in every project, this would be the first time he would be completely in charge…and could dive as much as he wanted. Divers shared 50/50 in the diamond production and he was convinced he could *strike it big* now that he had the experience. I gave him the supplies he requested, wished him well, and dropped him off at Imbambadai air strip.

A week later there was a call from Mrs. Mittlehauser. "You better come quick. One of your divers is here and there's an urgent message from B.G. Airways. It's about Frank. Something's happened to him."

Jorge, one of Frank's diving crew, was sitting on the steps of Mrs. M's boarding house. He came up to me and spoke fast and incoherently. "We don't know exactly what happened. The other divers were mad at Mr. Frank and when he didn't respond, they pulled him up. He was unconscious. We finally got him breathing but he's funny in the head. You better go quickly and get him."

"Whoa, slow down. Start from the beginning—from when I dropped you all off at Imbambadai."

"Well, we all were eager to start. On the second day we were diving, even though Mr. Frank complained that the camp wasn't properly set up. We don't care about that. We want to dive for diamonds. So Maiko, Jaime, and I started diving while Frank set up the tents. He was mad 'cause we didn't help. Maiko and Jaime are the best divers and by noon we were picking up diamonds. Soon as Mr. Frank come and hear that, he want to go down even though it was my turn next. He say he boss and he going down. Well, he move away from the big rock where they were working, started another hole and no diamonds. He come up and accuse them of lying. I go down and find first hole and the diamonds come up again. Mr. Frank even madder now and shut down dredge for the day. No talk in camp that night and no one tell Mr. Frank his dinner is good. We know what you told us, but we were mad, too.

In the morning we got late start and Mr. Frank go down first and work until noon. No diamonds. Then Maiko, Jaime, and I dive in the afternoon, but this time we only uncover diamond gravel but not suck it up until we ready to quit. Then we suck all the gravel around rock. Plenty good diamonds came up. Mr. Frank had problem. Didn't know if he was happy about diamonds or mad because he didn't find them. He good cook but not good diver.

Next two days same thing. Maiko and Jaime found good pockets of diamonds but they don't leave the dredge intake hose there. Mr. Frank go down and has to make his own hole but he don't know nothing about diving for diamonds. He gets nothing. And he gets madder and starts telling us what to do. With him diving we wait hours with nothing to show for. Then the problem got bad. The divers found another good pocket. Instead of working it, Mr. Frank say to shut down for lunch. No one shuts down for lunch when there are diamonds. Mr. Frank take keys for engine so we can't dive. When he come back later he say he take first dive but it not his turn." He hesitated to say more.

"So then what happened?" I asked. Jorge fidgeted but wouldn't look at me.

"Jorge, tell me what happened."

He said a lot that didn't make sense, but I finally got the crux of it. On a four man crew, one is the diver, the second tends the lines, and the third makes sure that the wind does not blow from the engine exhaust across the intake of the compressor supplying the diver. The fourth man cooks.

He continued. "Someone said," and he didn't know who, "Maybe Mr. Frank need a little change of air. Some of the engine exhaust went into the compressor—for too long. Mr. Frank didn't respond to tugs on the life line so they pulled him up. He was unconscious, and they were frightened they killed him. When they took the mouthpiece out of his mouth and rolled him over, he started breathing but wasn't awake. Other divers came over to see what happened. Mr. Frank sat there with his eyes closed and he wasn't moving. After a couple of hours, he wake up but was talking *silly* and saying the same thing over and over. The next day it was the same. They tell me to come to town, find you, and get Mr. Frank to a doctor."

Bertie drove us to the airport where I checked the plane, and Bertie retrieved the B. G. Airways message for me. It said basically the same without all the details—*Frank needs medical attention.*

We landed at Imbambadai. Jorge guided us on a thirty-minute walk to the camp. A couple of miners were hanging about and said that Maiko and Jaime stayed up all night with Frank. As soon as they heard the plane land and saw us coming, they took off. In the back of my mind I thought *and with all the diamonds, too.*

Frank was sitting on the cot and waved to us as we arrived.

"Hi, Don, what's up? What brings you here? Did something happen?" Then it went on, and on—the same questions over and over in a steady stream of words. Attempts at conversation were unsuccessful. He didn't know where he was, what he was doing here, and where he was going next...or why.

He obviously couldn't be left alone, so I brought him to Mrs. Mittlehauser's place and, bless her soul, she offered to keep an eye on him while I went out on business. The next morning I brought Frank to Dr. Derek Lyder's office and after a brief examination and discussion about what happened, he sat down with me and told me pretty much what I expected.

Frank had experienced carbon dioxide poisoning and/or lack of oxygen. Enough so that it affected the brain's short term memory. The unknown was how much damage there was, and if temporary or permanent. There was no way to know. Derek's diagnosis was that we would have to wait it out. There was nothing to prescribe or do at this time...only wait and see.

Mrs. Mittlehauser lasted four days. "I can't take it any more. He's driving me crazy. And worse, he's driving my ladies' groups crazy. My three bingo clubs come twice a week and they're all threatening to leave. Frank is driving everyone sheer mad."

This is all I need. I'm trying to find a seaplane and now I have Frank to nursemaid.

Doc Lyder recommended an East Indian boy, Rahaman, who sometime worked at the hospital but wasn't employed now. He needed a job. I hired him full time to keep an eye on Frank, and we put them up in the Spring Garden guesthouse. Mrs. Mittlehauser offered to check on him regularly and would dole out the money needed for the guest house, food, and agreed-to salary for Rahaman. That solved my immediate concern: returning to Wisconsin and the family over the holidays.

I kept in touch by cable with Mrs. M. She was doing a stellar job, and I would have to reward her handsomely for taking over *Project Frank*. Unfortunately, Frank did not improve quickly. Three months later at the start of the dry season there were a few signs of improvement but it was gradual—he would be lucid for an hour or two and then nothing the whole day. Progressively the lucid times expanded into half days. Another two months later Frank was back to normal with only occasional lapses that lasted a few minutes to a half-hour. Eventually even those disappeared. He had no recollection of what had taken place. He only remembered taking the dredge to Imbambadai but nothing more. I explained in detail countless times what happened. Even when his mind was clear he had constant questions about why it happened, were there any diamonds, and did I think it was an accident or on purpose.

I lay in bed thinking the price of diamonds goes up and up and for the strangest reasons. I thought of the people in the past who said, "Oh, diamonds don't have any real value. It's only the DeBeers Syndicate that keeps the price artificially high." I would like to have those people down here for one month for them to understand the real value. I had also come to the conclusion that no matter which direction we go, air charter, diamond trading or safari, it was necessary to have a seaplane.

25. FERRYING FLOATPLANE

With the help of friends, and particularly Glen Glendenning, an airplane buff, who had numerous contacts and a myriad of brochures on available planes, we found a seaplane that would fit our needs—one we could afford.

Glen helped us find a qualified seaplane pilot in St. Louis, and after many telephone calls and references from former employers, we arranged to have Lew Winters join our operation. Two weeks later, we picked him up in St. Louis and flew down to Kissimee, Florida to see the plane we were planning to buy. We completed a thorough inspection but had to accept the seller's word on the engine work. Two days later we were the proud owner of a Convair L13 on pontoons. The next few days were spent checking out the "bird" and making sure it was airworthy for the trip to South America. Besides being a good pilot, instructor, and helicopter pilot, Lew had all his A & E engineering degrees and was qualified to do all required work on aircraft. He could keep up our logs to the satisfaction of the FAA and the insurance company. It looked like we were off to a good start.

It took 2 months for Lew and Glen to ferry the Convair L113 floatplane from Florida to Grenada. By the time it arrived in Grenada, Jan and I both had the eerie feeling that this particular plane had a curse on it—a feeling which eventually proved itself true.

Donald Haack

First, there were the multitude of mechanical and engine problems in Florida that Lew worked on until he felt the plane was airworthy. The problems were exacerbated by the seller's personal assurance that the engine had been rebuilt in his own shop and everything was in first class shape. This, we found out later, couldn't have been further from fact. He said the *little* problems that Lew corrected were inconsequential because, "The Convair hadn't been flown much for the past couple of months."

The delays in Florida were not significant but were a prelude to what was going to happen. Less than four days after leaving Florida, our plane was impounded for seven weeks in Dominican Republic. Lew had engine problems two hours out of South Caicos, declared an emergency, and landed in Dominican Republic. Big mistake, but no other choice.

Small countries have a phobia about uninvited unannounced airplanes landing on their territory, emergency notwithstanding. Arriving without a flight plan is asking for trouble. To worsen matters, the Dominican police tore apart the aircraft, under the guise of a routine inspection, to discover an undeclared handgun on board.

The mechanical problem of dirty fuel in the line was corrected in a few hours. The matter of the gun took three weeks, diplomatic intervention, letters from the States, and copies of the Guyana Department of Civil Aviation showing that it was a requirement for every bush plane to carry a handgun, emergency rations, and medical supplies.

Five days after Lew and Glen's release, they landed our floatplane in the picturesque harbor of Lance Aux Epines. We could see the plane from Casa Miranda, the house Jan had moved into with our children. The Convair bobbing about in the bay was truly a stunning sight and conversation piece for the Calabash Hotel guests and the marina crowd. The bright yellow colors created a stark contrast to the clear sapphire blue waters of the Caribbean harbor. Glenn, the copilot, didn't want to continue and elected to stay on as an uninvited guest. Lew didn't think he could handle the plane alone, particularly the docking in a strange river in a foreign country. It was a long trip to Guyana, starting with 90 miles of open water from Grenada to Trinidad. The plane would be out of sight of land for a half hour. The route along the desolate and uninhabited eastern coast of South America included the 75-mile passage across the Orinoco Delta mouth, an area that Lew was neither familiar nor comfortable with. I agreed to go with him on this last leg.

We called Pearls Airport across the island and asked the Shell Oil Company to send out a 55-gallon drum of aviation fuel. We siphoned it into 5-gallon jerry cans and rowed them out to the plane. Handing up 45-pound cans of gas to Lew, standing on top of the wing, became a balancing act. He steadied himself as well as he could in the rolling swell and poured the gas into the funnel. We managed to pollute the pristine bay with a little slick of multicolor gas floating on the water. Eventually all tanks were filled.

The big day arrived. We had taken our boat out several times to check the wind and water conditions, which were tricky at best. The heavily loaded Convair needed a long run for take-off, the full length of the bay. The crosswind gave us little help to shorten that run.

Departure day was beautiful. Our neighbors in L'Anse Aux Epines, the Red Crab crowd, hotel guests, and the marina *yachties* came to wish us well, bid us farewell, and watch the "tricky" takeoff, as it became known in the local pub. Lew started the loud and throaty 300- horsepower Lycoming engine, shattering the peaceful quiet of the bay. We had to take off inside the bay where the swells were small. To be in the large swells of the open ocean wasn't an option. We taxied back and forth on short trial runs until Lew felt we had a good margin. As it turned out, it wasn't all that good.

Lew taxied to the innermost part of the bay, as close as possible to the shore in front of the Calabash hotel, turned into a slight crosswind, and pushed the throttle to full power. In two minutes we traveled three-quarters of the bay. The open ocean loomed uncomfortably close. Partially airborne and bouncing from wave to wave, I wondered how much pounding the floats could take before they collapsed. After one particularly hard bump, as we came onto open ocean, we were airborne and hung precariously a few feet above the water. It seemed forever. We were on *ground effect*, that ten or twelve feet between the wings and the water that trapped and compressed the air to give more lift to the plane. We hung there for the next hundred yards, gradually gaining airspeed and altitude. At 500 feet, we eased back in our seats. It was a tense takeoff.

The trip to and over Trinidad was uneventful and beautiful, and our engine had a comforting dependable sound. When we contacted Piarco Air Control, they routed us directly across Trinidad, thereby saving us precious fuel. An hour south of Trinidad, the crystalline blue of the Caribbean turned to a muddy brown, announcing the mouth of the huge Venezuelan Orinoco River. The silty-brown water extended twenty miles out. Eventually the sea reverted to its beautiful azure blue as we flew along the Guyana coast. From here on down, in the event of an engine problem, we had a security blanket—we could glide inland and land on the protected water of the Morawanna River, which paralleled the coast..

Georgetown revealed itself long before it could be seen. The Demerara River spewed forth the brown silt of the interior that spread like a huge aneurysm protruding from the continent. The depressing dark water created a Rorschach pattern that projected unpleasant images of events to come, dampening our buoyant spirits.

The Demerara River was just ahead. Georgetown, seven feet below sea level and void of tall buildings, was difficult to see from afar. Leave it to Dutchmen to build a town below sea level. I remembered from history the Dutch had traded Manhattan Island in the new world for the territory of Guyana. Not one of their best trades.

As we approached the air control zone 30 miles out, we prepared for the landing. Still in our swimming suits from the wet preflight and hauling up of anchors in Grenada, we opted to dress after we landed and secured the plane. We had no premonition of what was in store for us.

Donald Haack

It was my first trip in a sea plane. Lew briefed me on how to coil the ropes and set the anchor, if we didn't tie onto a dock. With our two throw-out lines and two anchors, he figured we had our bases covered.

Guyana Airways seaplane ramp, our designated landing area, was one mile in from the mouth of the Demerara. We called Atkinson tower, 30 miles upriver, to announce our arrival and advise that we would switch over to Guyana Airways frequency.

"Roger, Convair 205 Bravo, over to Guyana Airways. Please close your flight plan when you dock. Atkinson Radio"
His voice had the lilting quality, rising a few notes at the end of sentences, giving it an almost musical sound that was so peculiar to Guyanese. It had a charm all its own.

"Roger, switching to Guyana Airways frequency and will close flight plan upon docking. Bravo 205, over and out."

"Guyana Airways, this is Convair 205 Bravo, do you read me?"

"Roger, Convair 205 Bravo, loud and clear. We monitored you on the Atkinson frequency. Understand you are expecting to land at our ramp?" The voice sounded slightly incredulous. And this time it went way up several notes.

"Roger, Guyana Airways. We cabled your office last week and received permission to clear customs and refuel before going up river. Is that still okay? 205 Bravo"

"205 Bravo. Yes, yes, that's perfectly okay. I have the permission in front of me. Welcome to Guyana. But we only know that a Convair is 65,000 pounds with four engines and never heard of one on floats. All we got here is water. No land. Please advise. Guyana Airways."

"Guyana Airways. Okay. This Convair is a light plane, single engine, and does have floats. Your water will do just fine. Thanks for the invite. Will call you on final. 205 Bravo."

The light and humorous exchange probably broke the monotony of their otherwise boring day.

I didn't expect the number of ships in the area nor the white caps on the water. Lew circled the area. We descended to 400 feet.

It didn't get better. There were telltale trails of white foam coming from the seaward side of the docks and anchored ships, indicating a very heavy tide going out. It didn't look good to me. I expressed my concern to Lew and suggested we change plans and go upstream near the Atkinson airport to protected water without traffic.

"We built a temporary dock and could advise the tower to bring customs and immigration to meet us," I shouted over the engine.

Lew brushed the suggestion aside. "I've flown this baby all the way down from the U.S. in all kinds of weather and conditions. This is a piece of cake".

He chose a spot into the wind on the south side of the river, alongside two large ships anchored on the dock. The waves there were rolling but not whitecaps. We flew the downwind leg at 200 feet, turned onto the base leg of our landing pattern, and cranked in half flaps.

"Guyana Airways. 205 Bravo. Turning onto final, over."

"Roger, 205Bravo. Have you in sight. Take caution. There's a heavy current running and the tide's going out, over. " *That's an understatement.*

"Acknowledge. Thanks. 205 Bravo."

As we turned into the final leg, Lew adjusted the power a little higher so we hung at 50 feet above the water. When he was certain of his alignment, he eased up on the power and we settled in. We hit hard. He powered on for a few seconds and then eased back for second bounce. That one was better and we settled on the water. Lew reduced the power, pulled up on the nose, and the Convair L13 taxied toward the Guyana Airways' ramp. That was when we became aware of how horrendous the current really was. The combination of the river flowing out and the outgoing tide at its peak created the strongest current I had ever seen. Lew pushed the throttle to full power and still had difficulty maneuvering the Convair towards the dock.

"We're not getting a lot of rudder control," Lew yelled over the roar of the engine.

As we approached the ramp the current swung us away before Lew could maneuver close enough for me to throw a rope. He reapplied full power to steer clear of a ship behind us. We got out of there fast and turned 180 degrees for another try. At that point, I knew I was on the wrong side to throw a rope to the ramp. Lew tried taxiing into the small area closer to shore, but again the swift current swung us away from the ramp. This wasn't fun and we were in a very confined space with a plane that didn't have good rudder response.

Being on the far side, I couldn't throw a rope unless Lew got in quite close, which he couldn't seem to do. I felt helpless and climbed back in to ask Lew what he wanted me to do. He seemed too preoccupied to talk. All his attention was focused on trying to maneuver the seaplane into that small slot of water in front of the dock.

Then, inexplicably, shouting to me to throw out the anchor, he shut down the engine. I climbed back onto the float, opened the cargo door, pulled out the anchor, and heaved it out in front of the float, being careful the outgoing line did not tangle in my feet to pull me with it.

The line went out with a speed that took me by surprise. Lew saw it too.

"It's going out too fast. Get the line secured onto that cleat quickly or you'll never hold it. Secure it as tightly as you can. This current is unreal!" he shouted

I glanced back at the coil of rope and it was almost all out. I grabbed it a couple of feet back and strapped it around the cleat. The rope immediately grabbed hold, swinging us sideways to the current. I almost fell. I bent down, grabbed the strut, and put more rope turns on the cleat—the force of the current was actually pulling the rope around the cleat. This was unbelievable.

The shore was passing us at a fast rate. We drifted a couple hundred yards downstream from the ramp where Lew shut down the engine. The pull on the anchor was frightening but worse— the anchor was not holding in the mud. I shouted that to Lew, preoccupied now trying to re-start the engine. Before we touched down, I was glad to have that one-inch heavy rope and the 30 feet of 3/4 inch chain to set the anchor. Even that wasn't enough in this current. The anchor, barely slowing us down pulled through the mud.

"Stand back from the prop. I'm going to start the engine again. The anchor's not holding. Pull it up when I taxi back up stream."

He hit the starter for the second time for a longer duration. I wondered how long the battery could hold. That was a big engine. Not even a small firing. Not good.

"Throw out the second anchor. That may slow us down and give me more time to get this started," he shouted.

I threw out the second anchor, played out almost all the line so it had a better chance of setting, and secured the rope to the forward cleat.

I was so engrossed in the anchor operation, I hadn't noticed that even with the second anchor secured, we were drifting backward. We were literally plowing the anchors through the mud. Lew glanced back anxiously. I knelt down on the float and looked below the plane in the direction of Lew's attention. We were drifting into an ocean-going ship tied to the dock. Worse, tied alongside the ship and slightly behind its bow, was a huge seagoing barge. If we were lucky to miss the bow of the ship, we would hit the barge head-on and get sucked under. Now I could understand the panic in Lew's voice.

If the engine started, we could still maneuver to a more secure spot. But it didn't. The engine was turning but not firing. *This can't be happening.*

Lew again hit the starter for a much longer time. Another minute and the prop slowed down. The battery was spent. The six-foot prop spun around a few feet in front of me, but slower now, and nothing happened. He hit the starter again. It went through 30 seconds of revolutions and stopped.

"It's probably flooded, Lew. It's a hot engine. Shut it down. Try closing off the fuel, open the throttle wide, and hit it again."

He didn't respond. I opened the door wider and repeated what I just said. He wasn't listening. He was almost catatonic. I shouted again to him.

Donald Haack

"I know this engine. I can start it, but it's always a hard start. Throw out the second anchor or we'll be in real trouble."

"It's already out and it's hardly slowing us down." I shouted.

He pointed over his shoulder. The ship, directly behind us was less than a hundred feet away. Alongside it, the huge ocean going barge with its 70 foot wide flat bow, loomed ominously in our path, its deck slightly higher than our wings. A god-awful sight. It spelled catastrophe if we drifted into that. And that's exactly where we were headed.

Lew, you have to get the engine started. The second anchor isn't holding."

He turned the prop again without results. The battery was grinding down. From the sound of the starter, my hope for a start faded. Lew pressed hard on the starter button as if the extra force would help. Nothing. The prop barely turned.

I looked back. We were coming up on the ship, now only a few feet back. We drifted into it. The wing touched the hull. We bumped and slid alongside. I wondered how much damage was done. But now, the real danger—the barge. We were drifting right into it. I could hear the sucking sound of the water going under the slanted bow—a giant whirlpool sucking water underneath its hull. The barge appeared to be bearing down on us at a speed of 15 knots, when in reality it was standing still. The water was going under the barge's bow at 15 knots. And soon so would we.

After our wing hit the ship, we slid alongside. Lew stopped trying to start the engine. In this position, it would do no good even if it did start. We would ram the ship with the propeller.

Just before we touched the barge, I realized that saving the plane was out of the question. Up to that point, it was foremost on my mind, but now the chance of Lew and me getting out was almost as slim. As our wing scraped along the ship, I went to the front of the float to see the space between the ship and the barge. I might be able to dive in between the barge and the ship and swim to shore behind both of them. At that instant an ocean swell came and the barge slammed against the ship, flattening the bumpers along the way.

Donald Haack

I could be flattened the same way. It wasn't a nice picture, and I quickly drove it from my thoughts. There had to be another way.

I had checked the river side first but now it was my only option. The barge was a good 70 feet wide, which meant I would have to take a running start, dive off the back of the floats, go down deep, swim the 70 feet underwater, and hope not to get disoriented. I had to surface on the far side of the barge. I didn't know how loaded the barge was, but it looked like it was floating high with only a five or six foot draft. I had to give myself a good margin. If I were wrong, the current would drag me the 300 or 400 feet under the barge. Even if I could do the impossible and hold my breath, I'd be skinned alive by the sharp barnacles along the hull. I quickly put that thought out of my head, too.

I shouted to Lew, "Climb up on the wing and get on the barge. It'll be seconds before the plane breaks up and goes under."

He had the same thought. He was up and over. There wasn't time or space for me to try that. I was on the wrong side. With the plane slightly tilted from the force of the water, the wing was partially above the deck and secured by the deck hands on the barge. Someone threw Lew a rope to steady himself as he clambered up the wing. Confusion reigned—directions were shouted. The men on the barge understood the problem and tried to walk the wing away from the ship, along the front of the blunt bow, to push it off where the plane would be free to drift into the safety of the open river.

It was a great plan but never had a chance against the unrelenting current. Some deckhands grasped Lew and pulled him to safety. The plane continued its tilt, losing the battle with the forceful water battering its floats. As strong as they were, the men holding the wing faced a losing battle. That many deckhands weren't enough to move the plane sideways and keep it from further tilting. The river was winning. The sucking of the current, strong and unrelenting, was bending the wing. It would soon collapse.

The water sucking noise was unreal—like water going down a giant toilet bowl. I could no longer make out what everyone was shouting. The wing on my side came perilously close to touching the rushing water. When it did, it would be all over.

I hyperventilated. When I dived I planned to hit the water as far in front as I could, go deep, swim underwater far enough to clear the barge and into the safety of the unobstructed river.

My main concern was to go as deep as possible before I was sucked under the barge hull. The tilted plane made my short run on the pontoon more difficult. The wing was hovering a foot above the water now. I hyperventilated one last time. I needed every advantage.

I prepared for the short run on the float. The wing buckled. I knew it was over and the plane would be sucked down momentarily.

Then I heard it—or more correctly felt it. It was impossible not to—the fog-horn blast of the ship alongside. At sea, the horn blast can be heard miles away. This close, the sound reverberated throughout my body. Ships never sound horns at dock—it would shake the whole town. The vibrations hurt my ears and chest, and instinctively I looked up to the source. The captain, almost directly above me on the ship's rail, waved his arms and frantically pointed down to me.

A huge "S" hook from the ship's crane swung inches above my head. I grabbed it with both hands and nodded my head in an affirmative gesture. I rose a few feet above the plane. I heard a large slurp. The seaplane disappeared below the barge. Nothing but churning brown water below. I shuddered, clung tighter and bent my arms to ease the shock of any unusual jerk that might break my grip on the hook. When my feet hit the deck, the men had to pry my fingers loose. I watched the hook go up and there was the captain, with a thumb's-up sign. I saluted in thanks. He saved my life.

The world was topsy-turvy. I stood in a daze, barefoot and naked except for my swimsuit. Men came up with cheers that we were saved, condolences on the lost plane tragedy, and whatever else they said. It was all one big blur. I searched for Lew. He was sitting down, all alone, a few feet from the melee of workmen. His head was bent down to his knees and his arms were wrapped around his legs in a fetal position. He was shaking or sobbing, I couldn't tell which. I quickly went to him, knelt down, and asked if he was hurt. Eventually, he shook his head, no. If he was in shock, I had to get him to lie down, feet up and keep him warm. This apparently was a general breakdown of nerves. I waited a few minutes until the gasping and sobbing subsided and got him to his feet. We were helped up a ladder onto the ship, and later down onto the dock into a waiting taxi.

We arrived at Mrs. Mittlehauser's boarding house, rang the bell, and were invited in by an older lady who was taken aback at the sight of us. We entered in the midst of Mrs. M's afternoon tea party—twelve ladies sipping tea, discussing the day's gossip. To their credit, they hardly blinked.

Mrs. M, in her proper demeanor, welcomed me back and asked if we had a good trip.

"It's good to see you again, but no. We had a particularly bad trip. Tell you about it later."

In our bare feet and bathing suits we walked by the ladies in their afternoon finery and nodded to each as we passed. It's hard to be dignified when you're almost naked, but we put on a good show. They in turn responded well. I felt we gave them a bit of fresh material for their afternoon discussion.

The next morning, in spite of a good sleep and a hearty breakfast prepared by Mrs. Mittlehauser, we had a difficult time shaking off the depression of yesterday's disaster.

Before we had time to even contemplate the next move, there was a knock on the door. A stern-faced custom's officer handed me some papers. He expected immediate payment in full for the duty on the airplane we imported. My emotions were confusion mixed with murder.

I explained the circumstances in a calm a manner as I could muster, but he said he knew the details. No comments or expressions of sympathy. *The plane was imported and he wanted the duty.*

"Fill out this form. Describe the details of the plane here. On the bottom write in the fair market value."

Donald Haack

Of the two mixed emotions, murder was winning out. With much restraint, I asked, "Have you seen the plane? And where is it now?"

"That's not my affair. We have witnesses."

"You haven't told me, where is the plane now if it's imported into the country?

"I don't exactly know."

"Well, when you find it let me know. In the meantime please leave." I closed the door. I made the decision that boiling this son-of-a-bitch in oil would be much too humane. I would think of something more appropriate.

26. R & R AND NEW YORK INVESTORS

Afelr the sinking of the seaplane, I was badly in need of R & R. I sent Lew back to the States, put the operation on hold, and departed for Grenada. Jan, the kids, sunshine, sandy beaches, and rum punches might put a different spin on life.

John and Anne Pond, my favorite couple, and our daughter, Julie's godparents, were to join us in Grenada the next week. John was Vice President of Kidder Peabody in New York and one of the young up-and-coming Wall Street favorites. The job was high stress. He and Anne were in the final stages and finishing touches to the Darien, Connecticut dream home they were building. Both were in need of some good relaxation. Grenada was the medicine we all needed.

I arrived a week before the Ponds, and when I greeted them at the airport I realized they needed this week in Grenada much more than I did. Stepping down from Liat's shuttle airplane, Anne greeted us with her arm above her head. She said there was a painful nerve in her shoulder and the only respite from pain was to hold the arm on top of her head. Her several sessions with a physical therapist offered little improvement.

The image of Anne holding her arm above her head and eating *fish and chips* at the Red Crab Pub was hilarious. By the time dinner was finished, our sides were aching from laughter—a big stress reliever for Anne, and the first time in ten days she could actually bring her arm down for short periods without pain. Two more days of lolling in the ocean with rum punches, walking the beaches, telling stories, and at night playing cards—the sorriest Bridge imaginable—brought her arm back to normal. Grenada will do that.

We analyzed and post-mortemed John's high-stress job, and Anne's frustrations of building the Darien house, which was substantially over budget and five months behind schedule. Jan and I in turn poured out our sad tale of our lost seaplane. We exchanged Solomon-like judgments on what should and could have been done, now, and soon the world appeared rosier. Of course, the island rum punch played a significant role in this rehabilitation process. We pledged to have these soul-bearing, tension-releasing sessions more often.

Their last day was spent sipping our famous rum punches as we bobbed about in the shallow Caribbean blue waters of Grand Anse Bay. The sugar-white sand beach framed a picture-book setting. Small waves gently lifted and lowered us in the healing salt water, extracting any remnants of stress. The leaves of the sixty-foot tall coconut palms rustled with a soft whisper that surrounded us with background music. Our beach cookout of chickens, brats, and beach drinks completed the perfect tropical island atmosphere—an image forever etched into our memories as the height of good living.

As our jug of punch reached the halfway mark, John sat down beside me. "I'm part of a group of three who invest in small businesses. This is something we do outside our regular jobs—it gives us a break from the usual routine. We've done six start-up businesses already, and we're happy with the results. I'm interested in your air charter idea for the bush and would like to present it to Lacy and Herman, my other two partners. Not the safari and not the mining. Would you consider taking in partners? The flying seems to be the most lucrative business and the one you know best. What do you think?"

His offer took me completely by surprise. Business or any thoughts of business were a million miles away. I held up my glass so the sunlight could pour through, making little rainbows as I swirled it. My mind tried to grab all the racing thoughts on the proposition John had just presented.

"John, I'm flattered that you have that much confidence in me. Frankly, I've never had partners before, but this might be the right time. Let me ponder this a bit. I'd like to discuss it with Jan. Let's talk about it further over dinner tonight at the Calabash, a hotel owned by our friends, Aminge and Brian Thomas. I guarantee it will rank at or near your top dinner experiences. The atmosphere is super and we can easily talk there."

The Calabash was everything I promised and more. The after-dinner business discussion of the air charter was brief and positive. We seemed to be in accord on the basics.

Two weeks later I received a cable from John:

"Red Baron: *Air Charter* got the go-ahead from Lacy and Herman. Start the wheels for the license with your friend, the Minister of Communication. Have scheduled a meeting with Chemical Bank next Friday. Imperative you be there with a financial proforma operating plan. Confirm. Flying Ace, John."

Fortunately, I had a working financial *pro forma* based on one and two aircraft that could fill the country's needs for several years. It also would be easy to expand. I fine-tuned the figures and airmailed them to John the next day. The *pro forma* should strengthen John's presentation and give the bank a better financial picture of what to expect.

Before he left, John explained how their group of three operated, using a highly leveraged mode of financing and little of their own capital. Each of their businesses had separate corporations with signed personal notes against their considerable net worth.

I cabled back that I would meet him in New York the day of the bank meeting. I'd overnight in Antigua and board the early Pan American flight to New York. That would give me a couple of hours to clean up and change clothes before the meeting.

John sent a driver from Kidder Peabody to pick me up at the airport. He took me directly to KP's office on Madison Avenue. My first thought: I could clean up and change clothes in John's quarters before the meeting. Jan made sure I had my best suit, a charcoal grey, and I would blend in with the Wall Street crowd. John had other ideas.

"I'm glad you're early. The Chemical Bank people called and said to come as early as possible. Give us a chance to talk informally before lunch. Let's go."

"John, I can't possibly go to the executive dining room wearing what I have on (corduroy trousers, casual boating shoes, and a green sweater with leather patches on the elbows). It'll look like I was dragged right out of the jungle."

The first response I got from him was a huge smile. "You're dressed just fine. They know you're coming straight from British Guiana and the jungle, so it won't make any difference, and you don't have to apologize. Lacy, one of my partners, is joining us"

We arrived at Chemical Banks's top floor executive restaurant and were greeted by Lacy, two middle aged gentlemen, and a younger man about John's and my vintage. They gave us a great welcome. Drink orders were taken as we were ushered into a semi-circle of plush leather armchairs in a private dining area. They handed me their cards—a Senior Vice President, the Chief Executive Officer and the youngest was Vice President in Charge of Special Projects. Impressive. *John and his boys go right to the top and don't play around with middle management.* The martinis arrived. The next few minutes were spent on pleasantries, my trip up, the weather contrast, and other mundane subjects.

John leaned over. "I was telling Gene here," and he pointed to the CEO, "about some of your experiences. Tell them what happened when you were helping the Brazilian priest on a mercy mission and were put in jail and house arrest."

I tried to give a short version but with the barrage of questions, the misadventure was described in minute detail. Prompted by John, there were other questions that ended in long anecdotes. Lunch came. The questions continued unabated. At two o'clock, the three bankers stood up and expressed thanks for an entertaining and enlightening luncheon. We exchanged pleasantries and they left. Lacy joined them. John and I remained at the table.

"John, I can't believe this. I came prepared to give a detailed financial analysis of the air charter. I knew it from front to back—short, medium and detailed versions for any questions they would throw at me. The air charter wasn't even brought up. Now what, do we have to schedule another session?"

John just smiled and put his hand on my arm in a patronizing manner that a parent might use with a child. I was frowning. Lacy returned to join us. He nodded to John and whispered in his ear. John turned to me.

"Okay. It's all done. They're depositing $75,000 in our account tomorrow morning. You and I are flying up to the Helio Courier Company in Bedford, Massachusetts. Yesterday, I talked to their president, Gus Bullock, and his regional manager, Hunter Blackwell. They're expecting us and have a plane with all the specs we requested. Chemical's group was pretty well convinced from the *pro forma* you sent to me, which I passed on to them. But they wanted to meet and press the flesh of the real bush pilot and hear some of your exploits. They're going to live this experience vicariously through you and us. They actually agreed to a credit line of a quarter million, but we'll only draw on it as we show a consistent record before we buy more planes."

I could only shake my head and mutter, "This must be the difference between small companies and big ones. Haven't seen anything like this before." I understood why John didn't want me dressed up in a suit. As a bonafide bush pilot, I had to look the part—as if I just stepped out of the jungle. I *was* the pony show.

I just experienced *big* companies in action, and I liked what I saw. Later in the game, when I saw the downside of *big* companies and how they operated, I wouldn't be so enthused.

The next morning we flew up to the Helio Courier operation in Bedford and met with Gus Bullock and Hunter Blackwell, who gave us an extensive tour of their facility. It was impressive. They had five planes sold, each in different stages of production. We checked out their demonstrator that was in pristine condition and available to buy. Because it had 100 hours on the tachometer, they discounted it twenty percent. The paint job was a beautiful design in blue and grey. We were hooked.

After our discussions of how and where we were going to use the plane, the Helio technicians removed the more sophisticated navigational and radio gear. Guyana had none of this equipment, and it would only add weight and dollars to the plane. We flew a thirty- minute check flight with their test pilot, who demonstrated the short takeoff and landing capabilities. When the airspeed dropped to 45 miles per hour, the leading edges of the wings popped out, allowing the plane to fly even slower. It settled in like a helicopter. I was impressed.

John handed Bullock a check and bid us farewell. I spent the next couple of hours learning the idiosyncrasies of this unusual airplane.

The company displayed photos of planes in numerous bush operations throughout the world and were proud of the fact that the airplane's hull was built like a tank—not one recorded Helio fatality. One impressive series of shots showed what happened when one of their planes flew into a mountain. Both wings were sheared off and the shell tumbled a thousand feet down the slope. All four passengers survived with minor injuries.

Two days later with a fresh set of Caribbean maps and three hours of hands-on training in this new bird, I was winging my way down the eastern coast of the States and the Windward Islands to Guyana. It was the first time I flew solo on this route. I found the solitude that I needed before starting out on our new venture. The beautiful, easy and uneventful trip bode well for a new beginning.

I landed in Grenada which gave Jan and me the time to rethink and regroup. The safari was never meant to be a permanent part of our plans but it provided a substantial income and opened the door to another business enterprise; the air charter service so desperately needed in Guyana. The need plus my expertise and years of flying experience pointed to this logical direction. We could hire pilots and wouldn't have to live in the remote bush, as we did with the mining and trading company. Jan and the children elected to live in Georgetown until the office and air charter company was established and then return to Grenada to build our home.

A week after I arrived in Guyana, Minister of Communications Eugene Correia, issued our company a temporary permit. Our first charters were booked and *Guyana Wings* was born. Three months later we'd add our second plane, a Canadian DeHavilland Beaver on floats.

I had a good feeling about this new partnership company. I was no longer *going it alone.* I'd learned several tough lessons from the bush—all of which served to strengthen my resolve to make a good life in this South American land of opportunity.

EPILOGUE

A question I hear often at my lectures and book signings is, "Will there be a third book coming about Guyana?" The answer is yes, the third of the trilogy, *Diamonds 'neath My Wings*, will describe the bush flying of our company, *Guyana Wings, Ltd.*, the trials and tribulations encountered supplying trade goods, rations and miners to the rugged interior. The following is a partial excerpt from one of the incidents encountered in the not-so-routine flying.

Diamonds 'neath My Wings

As is usual during the dry season, the morning sky was clear with a trace of ground fog that the sun would burn off before takeoff. Cominco, a subsidiary of Canadian Pacific Railroad, was our good customer and had two DC-3 loads going into Annai. The company needed Guyana Wings to carry them from there into the southern savannah. Both of our pilots had exceeded their flying time for the month so I took the charter.

Cominco had arranged for four Amerindians to help load our aircraft. I decided that two of them would come with me on the first trip to help unload and since they'd need river transportation at the new location, on the last trip in I'd carry the canoe tied to the float struts. I had never flown with canoes before, but I had seen photos of Canadian Beavers transporting them. Pierre, Cominco's helicopter pilot, was there to quell any doubts I had. He gave me a detailed map of where the camp was located. Since the canoe would create quite a drag, I decided to take only a half-cargo load on that last trip.

Twenty minutes after my first shuttle I recognized the area cleared for the helicopter. I was supposed to land as close as possible in the nearby river. After I spotted a shallow bend in the river, I flew two low-level passes searching for an area without rocks, fast shallow water, or logs that could flip the seaplane upside down. The river was relatively narrow, which meant the clearing between the trees was marginal, and I either had to find a wide enough area or a place where I could slip beneath the canopy and stop short. On the second pass I was confident with the chosen area and eased onto the water. The actual landing site was better than it appeared from above and close to their campsite.

We expected to get in five loads Monday but we were lucky to get in four. There was no partying that night. We were all exhausted from dropping and pulling up anchors, paddling the heavy Beaver, loading and unloading 2,000 lbs of cargo from a makeshift ramp that was moving and onto a plane that was moving, and then the reverse to unload. We all slept soundly and anticipated a hard second day. Originally that would have left only four trips but in order to catch up we had to do five. As usual, when on a tight schedule, the weather never cooperates: the fog didn't break until ten, forcing an even tighter schedule. We skipped lunch. I had peanuts and a banana while flying the second trip.

At 4:30 all the personnel were in except the last two Amerindians and Pierre who'd stayed behind to fly them in the helicopter. All I had on board the Beaver were some basic rations— and one canoe. I double-checked the aircraft and supervised the placing of the canoe and the securing of the back straps. I topped up the gas tanks, checked the fuel drains twice to ensure there was no contaminated water in the tanks. We finished at the same time. I cranked the engine, they shoved me off. The wind died down and I chose a downwind takeoff in the direction I would head. That would save me a few minutes. Seconds later I was airborne. This was a bit of a tight schedule. I'd have twenty minutes before sunset to unload, tie up the plane and refuel. I intended to overnight at their camp but fully expected that I would have to do an instrument takeoff through the ground fog in the morning, if I were going to be back in Georgetown for our scheduled flights.

It was a long hard day. I knew I was tired. I forced myself to double check everything: exact compass heading, cylinder head and oil temperatures. I glanced outside repeatedly for emergency landing spots should that become necessary. Fifteen minutes after sundown there'd be total darkness. Everything seemed to be in order and keeping the schedule was doable.

Then it happened—a tremendous bang that shook the airplane as if I'd hit something midair, or blew out an engine. I couldn't see anything out the windows. I immediately pulled pack on the power to reduce stress as I checked all the instruments. *Nothing amiss there.* The flight controls all responded. Totally mystified, I searched for a place to land and found a stretch in the river ahead. You simply don't fly into something midair without some major structural damage, yet a visual check revealed nothing…I could see no damage. I flew at 50% power, slowly descending, wondering what next when a second louder bang sounded, this time shaking the plane visibly. It seemed to come from the left side of the plane and again I visually scanned the wing, struts, and then down below to the canoe. What I saw put me into temporary shock. The canoe was no longer hugging the float strut but had pulled away almost a foot, creating a wind drag that I could feel as the plane veered slightly left. The knot was undone and the only thing securing the front of the canoe was the friction of the four or five wraps of the wet rope—slowly uncoiling.

A sixteen-foot canoe loose at the front and tied securely at the back would whip around and demolish the tail. Or, with its enormous drag would literally pull the airplane out of the sky. Aerodynamically there is no way in the world the plane could fly with that kind of drag. The sun was almost horizontal in the sky and the light shone in at right angles into the cockpit. My lower arms were covered with beads of perspiration producing tiny prisms in the sun, which added an unreal dimension to this crisis.

I instinctively kicked right rudder as hard as I could, skewing the plane to the right in an unnatural flight, but as the airflow hit the left side of the canoe it slammed back onto the strut. The noise and shock were more severe than before. There was no way I could allow the canoe to pull away again but the challenge would be to keep the plane airborne in this totally unusual configuration. It was not built to fly slipping sideways through the sky.

Donald Haack